URANIUM
WARS

URANIUM
WARS

THE SCIENTIFIC RIVALRY THAT
CREATED THE NUCLEAR AGE

AMIR D.
ACZEL

palgrave
macmillan

First published in 2009 by PALGRAVE MACMILLAN® in the United
States—a division of St. Martin's Press LLC, 175 Fifth Avenue, New York,
NY 10010.

Where this book is distributed in the UK, Europe and the rest of the world,
this is by Palgrave Macmillan, a division of Macmillan Publishers Limited,
registered in England, company number 785998, of Houndmills,
Basingstoke, Hampshire RG21 6XS.

Palgrave Macmillan is the global academic imprint of the above companies
and has companies and representatives throughout the world.

Palgrave® and Macmillan® are registered trademarks in the United States,
the United Kingdom, Europe and other countries.

ISBN: 978-0-230-61374-4

Library of Congress Cataloging-in-Publication Data
Aczel, Amir D.
    Uranium wars : the scientific rivalry that created the nuclear age /
by Amir D. Aczel.
        p.   cm.
    Includes bibliographical references and index.
    ISBN 978-0-230-61374-4
    1. Nuclear weapons—Research—History—20th century.   2. Nuclear
physics—Research—History—20th century.   3. Nuclear energy—
Research—History—20th century.   4. Science and state—History—20th
century.   5. Uranium as fuel—History—20th century.   I. Title.
QC773.A1A28   2009
621.4809—dc22

                                                        2009007276

A catalogue record of the book is available from the British Library.

Design by Letra Libre

First edition: September, 2009
10 9 8 7 6 5 4 3 2 1
Printed in the United States of America.

For Debra

CONTENTS

Eight pages of photographs appear between pages 130 and 131.

ACKNOWLEDGMENTS

This work was partially supported by a grant-in-aid from the Friends of the Center for History of Physics, American Institute of Physics, and I am very pleased to acknowledge this help with gratitude. I am grateful to the American Institute of Physics' Niels Bohr Library and Archive, its former director, Spencer Weart, its new director, Gregory Good, and its librarians—especially Melanie Brown, Scott Prouty, Jennifer Sullivan, and Julie Gass—for the help they gave me on my research stay at the Niels Bohr Library in November 2008.

I thank the administration of the National Security Archive at George Washington University for permission to reprint and quote from many declassified "Top Secret" and "Ultra Top Secret" documents in the archive.

I am grateful to the Niels Bohr Archive in Copenhagen for permission to quote from Bohr's unsent letters to Heisenberg.

I thank Professor Alfred I. Tauber, director of the Center for the Philosophy and History of Science at Boston University, for his continuing support and interest in my research. I thank the center's personnel, and the librarians at Boston University, for their help in my work researching this book.

Many thanks to my friend and agent, Albert Zuckerman, for his brilliant ideas that shaped this book, and I am indebted to his staff at

Writers House in New York for their help. I thank my editor, Luba Ostashevsky, for her enthusiasm for this project and for guiding it to completion. I am grateful to the tireless Alan Bradshaw who helped immensely with all aspects of turning this story into a book. My thanks also go to Roberta Scheer and Jen Simington for their work on editing the book.

PREFACE

Hardly a day goes by without a major news report about nuclear issues, whether it's the international community's response to Iran's nuclear program or the future of Pakistan's atomic arsenal. At the same time, some politicians and scientists envision a future in which nuclear reactors dot the country, generating electricity that will help break our dependence on fossil fuels. Nuclear energy can help us combat global warming because this power source does not entail the release of carbon into the atmosphere. But the promise of a carbon-free energy source is checked by concerns about the ill effects of nuclear waste, as well as the danger of another disaster like the 1986 meltdown of a nuclear plant in Chernobyl, Ukraine, the human toll of which we have yet to fully quantify.

While we are inundated daily by these news reports, few people understand what the information means: What do 9,000 centrifuges working around the clock at an Iranian nuclear center actually do? What is refined uranium, and how do these machines produce it? And what *is* the power that resides inside the nucleus of uranium, a humble silvery element found at various locations around the globe, and how does this element deliver the immense destructive power of a nuclear bomb?

Most people know about the atom bombs that destroyed Hiroshima and Nagasaki at the end of World War II, and many know that the operation that brought these bombs into existence was called the Manhattan Project. But few people know the full story that led to that massive endeavor: How uranium was discovered, how its properties were investigated, and how a fierce competition among several groups of scientists in different countries brought us a deeper understanding of uranium. Few people know that it is the atom of the element uranium that undergoes the unusual process of *fission*—by simply splitting in two when it is hit by a tiny subatomic particle.

Researchers, working frantically in a race to understand this process, found that splitting apart the uranium atom releases energy—as predicted by Albert Einstein's famous equation $E = mc^2$. Some time later, an even more intriguing possibility was raised—and soon it too became a reality—the possibility of creating a chain reaction. When a very large number of atoms of uranium undergo fission in a chain reaction it creates a nuclear explosion. Equally, when the chain reaction is kept under control, when it is more moderate than an explosion, uranium can create the energy generated in a civilian nuclear power plant.

We are now at the cusp of developments in world affairs for which our policies concerning energy, economics, and national security depend, in part, on a thorough understanding of the properties and uses of uranium. This is why the story of uranium is so important.

Our knowledge about nuclear processes, nuclear bombs, and nuclear energy originated in the Second World War. Scientists who escaped the brutality of the Nazis built the atom bomb in America, just ahead of their counterparts in the Third Reich. Nuclear power went on to have a controversial life—holding the potential to provide civilian populations with energy yet threatening the world as rogue states such as Iran and North Korea have used it as a subterfuge to defy the international community and continue to develop atomic weapons.

The destructive power of nuclear arms is our postwar legacy. Before the war nuclear processes were a scientific wonder, a mystery that researchers worked night and day to solve. *Uranium Wars* charts the lives and work of the scientists who brought us the knowledge to make an atomic bomb, evaluates their responsibility, and explores their triumphs as well as their failure to stop the bomb from being used against thousands of civilians in Japan, a nation then close to surrender. It is also about the interplay between a country's leadership and its scientific community. All of these are issues I feel most passionately about, questions that have fascinated and tormented me throughout my life.

In the 1970s, I was a student of mathematics and physics at the University of California at Berkeley, and I worked in a laboratory with radioactive elements, using the techniques developed by some of the scientists whose stories are told in this book. Through my study of physics, I had the privilege of meeting one of the major players in the drama of modern physics and nuclear discovery: the German physicist and quantum pioneer Werner Heisenberg. This meeting changed and redirected much of my thinking. As a young student, I was taken by Heisenberg's genius and his brilliant explanations of the quantum theory.

While Heisenberg never talked about his wartime work on atomic development in Germany, I knew that there was a hidden side to the life of this charming man. And more than two decades later, in the 1990s, evidence surfaced that Heisenberg played a significant role in the Nazi effort to build an atomic bomb. I became obsessed with the promise and danger in science and with the ways governments can manipulate scientists to do their bidding.

Many other scientists played crucial roles in the development of atomic power and weapons. Some of them knew exactly what they were doing and entertained no illusions about what governments might do with their work. Others were perhaps more naïve, or were willing to believe that they would have a say in the political decisions. The story told in this book is a complicated and fascinating tale about how scientists decoded a mystery of nature—in fierce competition with each other—and how their discoveries enabled the launching of the most enormous weapons research and production project ever undertaken: the Manhattan Project, which brought us the atomic bomb.

The book follows the scientific adventures of the men and women who played key roles in the great endeavor of learning the secrets of uranium, breakthroughs that led them to discover the processes of fission and chain reaction—the essential elements of both nuclear power generation and nuclear bombs. These scientists included the tireless and fiercely ambitious Lise Meitner, the Austrian physicist who throughout her life had to struggle against sexism and anti-Semitism and yet triumphed to become the first scientist to decipher the strange process of fission. Our story follows the innovative experiments designed by one of the most versatile scientists of the twentieth century, the Italian physicist Enrico Fermi, who thought he had discovered the production of transuranium elements in his lab when in fact he had made much greater discoveries about radioactivity and the nature of the atom. These would lead to his creation of a chain reaction under the football field of the University of Chicago. And we meet the eminent Danish physicist Niels Bohr, whose work on uranium fission was of paramount importance and who influenced the careers of virtually all the scientists involved in this voyage of discovery. The book describes the struggles, challenges, and triumphs of the scientists who worked on uranium, as well as their conflicts. Their collective work resulted in the atomic bombings of Hiroshima and Nagasaki, the Cold War, and our present nuclear age—with its great challenges brought on by the proliferation of nuclear weapons and the expansion of nuclear power as a response to global warming.

CAST OF CHARACTERS

MAIN CHARACTERS

Enrico Fermi   Italian physicist and Nobel Prize winner who immigrated to America, became the world's expert on neutron radiation, and created the first fission chain reaction at the University of Chicago in 1942.

Werner Heisenberg   German physicist and quantum pioneer, this Nobel laureate was involved in the unsuccessful Nazi project to produce an atomic bomb.

Lise Meitner   Austrian Jewish physicist who immigrated to Sweden, derived the theory of fission, and conducted groundbreaking work in physics.

Otto Hahn   German chemist and Meitner's collaborator, another Nobel Prize winner who remained in Germany, perhaps secretly opposing Hitler.

Irène Joliot-Curie   One of Marie and Pierre Curie's daughters, conducted key work on uranium processes, won a Nobel Prize, and was Meitner and Hahn's fierce competitor.

Niels Bohr — Danish physicist, Nobel laureate, developed a model of the atom and did theoretical work on fission; connected many scientists through his institute in Copenhagen.

SECONDARY CHARACTERS

A. Henri Becquerel — French physicist, discovered uranium radiation, shared a Nobel Prize with Marie and Pierre Curie.

James Chadwick — English physicist, Nobelist, discovered the neutron.

Marie Curie — Polish-French physicist and radioactivity expert, discovered polonium and radium, twice a Nobel winner in different fields.

Pierre Curie — Marie's husband and co-worker, shared her first Nobel.

Albert Einstein — His equation $E=mc^2$ made it all possible.

Otto Frisch — Lise Meitner's nephew and fellow physicist, helped his aunt derive the theory of fission, worked on the Manhattan Project.

Frédéric Joliot — Irène Curie's husband and co-worker, shared a Nobel with his wife.

Martin Klaproth — German chemist who discovered uranium.

Paul Langevin — French physicist, Marie Curie's colleague and devoted friend.

Ettore Majorana — Italian physicist and colleague of Enrico Fermi, disappeared mysteriously in

	1938 before the race for the atom bomb began.
Eugène Péligot	French chemist who further refined uranium.
Wilhelm Röntgen	German physicist who won a Nobel for his discovery of X rays.
Ernest Rutherford	British physicist and Nobel recipient who was first to identify an atom's nucleus and its particles through his work on uranium radiation, he mentored many first-generation nuclear scientists through his laboratory.
Leo Szilard	Hungarian-born American physicist who understood that uranium can create a chain reaction, campaigned against weapons proliferation.
John Wheeler	American physicist, codiscoverer of fission.

OTHER CHARACTERS

The story told in these pages, one of the most stunning tales in the history of science, spans many years and concerns the work of many people. Deciding who should be the main characters in the story was a delicate task that admittedly involved personal choice. Some people, perhaps with important roles in the drama played out in twentieth-century physics, are not listed above although they appear in the story. Others are not mentioned at all in the book. The work of these individuals and their influence on developments may have been equally important—but they did not fit the main thrust of the story I wanted to tell. For example, for reasons of brevity and emphasis I omitted many individuals who worked on the Manhattan Project, the culmination of the story of uranium. That project involved a large number of people, and I simply had to leave out all but a few of them if the story was to remain clear and tractable. Edward Teller and John von

Neumann, for example, played key roles in the American development of the atom bomb, but their work was not central to my theme; as a result, even though they were important, Teller and von Neumann are mentioned only briefly.

The list of characters above includes only scientists. The story, however, involves nonscientists as well, most notably leaders and politicians, military people, and diplomats. The key nonscientists in this story are:

General Leslie Groves	Military head of the Manhattan Project.
Franklin Delano Roosevelt	President of the United States, died in office on April 12, 1945, before the end of World War II.
Henry Stimson	Secretary of War in the Truman administration.
Harry Truman	President of the United States, was Roosevelt's vice president, became president upon FDR's death.

A NOTE ON NOMENCLATURE

I generally use the term *atom bomb* or *atomic bomb* to denote a bomb based on nuclear fission. This is not exactly correct from a descriptive point of view, because technically it is the nucleus of the atom—its very core—that splits in a nuclear fission inside the bomb, and not the whole atom (which is composed of the nucleus surrounded by the electrons). Still, historically this was the term used, and that is why I chose to go with this terminology. Also, I sometimes use *atomic physics* instead of *nuclear physics,* which is similarly imprecise terminology: atomic physics should be the term for the physics of the behavior of collections of atoms, such as the statistical behavior of atoms in a gas, while nuclear physics is the study of nuclear processes. I mostly reserve the term *nuclear bomb* for the hydrogen bomb, which is a later invention: a bomb based on the much more highly energy-releasing process called fusion, rather than fission, which is triggered by an atomic bomb as its core.

GLOSSARY OF ATOMIC TERMS

Matter   All the stuff of everyday life we see around on Earth: all solids, gases, and liquids in our environment or in space. Matter in the universe is composed of molecules, which may be groups of atoms (such as water, with two hydrogen atoms and one oxygen atom held in a chemical bond), or single atoms (such as helium gas, which is made of single helium atoms). There is also unaccounted-for matter—matter we know exists but we cannot perceive; it is called "dark matter" and, as yet, we know nothing about it.

Atom   The atom is the largest particle in this story (since we will generally not deal with molecules, which are in the realm of chemistry); its size is much smaller than can be seen. It has a nucleus surrounded by electrons.

Nucleus   The nucleus is the heart of the atom—the extremely small and heavy core at the center of the relatively huge atom.

Proton   The proton is a positively charged particle.

Neutron   The neutron is a neutral particle, almost identical to the proton but without electric charge. Both protons and neutrons are called nucleons because they live inside the nucleus.

Electron	The electron is a tiny, negatively charged particle.
Alpha Particle	The alpha particle is actually composed of four parts: two protons and two neutrons living together. It has the same composition as a helium nucleus (meaning a helium atom without its electrons). It emanates from the nucleus and flies out of it at great speed; it is a form of radioactivity.
Beta Particle	The beta is an electron that emanates from the nucleus, rather than from an electron orbital in the atom (i.e., revolving around the nucleus). It flies out of the nucleus at great speed; it is a form of radioactivity.
Gamma Ray	Gamma radiation is a third kind of radioactivity; it is a massless high-energy wave.
Neutrino	The neutrino is tiny but has a mass; it almost doesn't interact with matter.
Anti-neutrino	The anti-neutrino is the opposite twin of the neutrino.
Positron	The positron is the opposite twin of the electron; it has a positive (instead of negative) charge.
Anti-proton	The anti-proton is the opposite twin of the proton; it has a negative (rather than positive) charge.

INTRODUCTION

THE BLINDING LIGHT

August 6, 1945, started as a clear, hot summer day in Hiroshima, a city on the agriculturally rich delta of the River Ota in the southwest part of Japan's Honshu Island. At 8:15 in the morning sirens sounded throughout the city after apparent enemy planes were sighted high in the sky. Then a blinding light was seen, often later described as a huge bolt of light. Some survivors reported a series of flashes, followed by a deafening blast and then by powerful, scorching winds that would not ease and burned the skin and flesh of everyone in their wake. Within minutes, huge fires raged throughout the city. Hiroshima was a mass of burning flesh, metal, and wood.

The culprit was the atom bomb—aimed for the first time at unsuspecting civilians, a city of 350,000 people. The bomb, made of a rare isotope of uranium called uranium 235, which had been refined and purified from ore over the course of two years in a secret operation known as the Manhattan Project, was dropped over the center of Hiroshima from an American warplane named the *Enola Gay*. The nation's most advanced strategic bomber had been constructed expressly for this mission. Fifteen such planes had been equipped to carry atomic bombs for this and possibly subsequent operations.

In the early hours of August 6, 1945, the *Enola Gay* had taken off from the large U.S. airbase on Tinian Island in the Marianas, accompanied by two other B–29s, one of which was to take pictures of the bombing. After about 6 hours of flight, they arrived over Hiroshima, at 8:15 AM, at an altitude of 32,000 feet. When the *Enola Gay* was over the center of the city, Colonel Tibbets gave the order and the atomic bomb, nicknamed "Little Boy," was dropped.

The *Enola Gay* quickly reversed direction and sped away to avoid radiation damage to its crew. Little Boy fell through the air for almost a minute, and when it reached an altitude of 1,900 feet, by design its fuse detonated, setting off a small conventional explosive within the large bomb, which sent one piece of uranium 235 to meet a second piece, and nest perfectly inside it. When this happened, the total combined mass of uranium now located in one place exceeded the minimum necessary to undergo fission. The spontaneous fission—or splitting in half—of untold numbers of uranium atoms in a chain reaction caused the horrendous explosion. A tiny amount of mass was thus converted into an immense amount of energy, which created the huge blast. The explosion destroyed the city—and it launched the nuclear age in which we live.

People who were within a mile of the detonation area were completely vaporized. In one case, a person left us his shadow, etched through the intense radiation on the remnants of a wall. All buildings within a radius of a mile of the blast were pulverized by the explosion.

Setsuko Nishimoto recalled what happened. She lived in a village several miles outside the city. Her husband didn't want to go to work that day. Reluctantly, he joined his fellow villagers who left on oxcarts for Hiroshima, where they were employed in demolishing an old building.[1]

"I was in the lavatory of my house when it happened," she recalled, "I thought it was a flash of lightning, and next there was a noise, an enormous *Bang!* The house went pitch dark. The sliding doors and screens went down, there was an enormous blast of wind, and the wall fell right out. When I looked over toward Hiroshima, I saw a black cloud rising up."[2]

All around Hiroshima, Setsuko saw flames rising. It seemed that the whole city was on fire. She worried about her husband, but—not comprehending the immensity of what she had just witnessed—assumed that he had been redirected with his work crew to put out the flames. In the afternoon a person with a loudspeaker moved through the village announcing: "Hiroshima is completely destroyed!" During the night survivors of the blast were evacuated to a factory and tended to by medical crews.

Setsuko went to search for her husband. The "huge crowd of people," she recalled later, "were charred and terrible to watch."[3] People had most of their clothes torn and burned off their bodies. Their faces were so swollen that their eyes could not be seen. Their hands and feet were puffy from fire and radiation burns. Another woman described seeing a man who was so badly burned that "his skin looked like cellophane and was hanging off."[4]

Setsuko did not find her husband. But even for those who did manage to find their loved ones, there was no happy ending. All succumbed to the effects of radiation and died. A week later Setsuko burned with a 105-degree fever. Her hair fell out at the slightest touch. She had a severe case of radiation poisoning, as did many others who were not at the center of the explosion but whom the radioactivity nevertheless reached. Some survived but suffered from agonizing pain throughout their remaining years.

It is estimated that the bomb incinerated nearly 150,000 individuals in Hiroshima and at least 100,000 more died from radiation damage.[5]

Three days after the bombing of Hiroshima, the United States dropped a second atomic bomb, codenamed "Fat Man," on the Japanese city of Nagasaki. Fat Man was bigger than the first bomb and had a plutonium core. In this second attack 75,000 people were killed and many more died from radiation sickness and cancer over subsequent years.

The incidence of cancer in Hiroshima and Nagasaki has been shown to be strongly and directly correlated with the total amount of radiation absorbed by the residents of these two cities, and to rise sharply among people who had been closer to the location of the explosions.[6] As the novelist Kenzuboro Oe has said, surviving victims of the atomic bombs carry "the terrible burden of learning to live with their illness and preparing to die."[7]

Hiroshima and Nagasaki showed the world the kind of devastation that science can wreak—a complex device, delivered by plane or missile, capable of wiping out an entire city. The weapon was the result of a huge scientific leap forward both in its design and its development. It marked a drastic divergence from the conventional, chemically based bombs that had been used until that time.

What led to this terrible outcome? What preceded the devastation of these two Japanese cities? What was the role of science? How did an ancient coloring agent—uranium mineral—which for centuries had seemed benign, become the cause of such immense destruction? What led to its transformation into an agent of uncontrolled explosive power?

There have been other books that covered the building of the atom bomb. There have also been books that scrutinized the decision to drop the bomb. This is a different book. My goal here is to make accessible to the reader the science behind this singular development. Additionally, most books about the atom bomb were written at the height of the Cold War, when we faced the Soviet empire in a game of deterrence. Whoever had the bigger bomb, went the logic, could scare off the other. Now that we have moved beyond the Cold War, we can think of nuclear power differently—not as a wholly destructive agent but as a source of energy that some day could be made safe, which may satisfy our growing thirst for industrial, commercial, and household power while helping protect our planet from overheating. We should also learn to control the present phase of nuclear weapons proliferation—now that the old Soviet threat has almost disappeared—and ensure that the specter of nuclear holocaust is forever behind us.

Nuclear fission has a complex and interesting history. What were the teams of scientists who discovered radioactivity and atomic fission seeking? How did scientists first come up with the idea that an atom can be split, producing a large amount of energy? What could possibly have led researchers and thinkers to assume that the atom was not an immutable piece of matter, solid as a rock, but rather something malleable that can, under the right conditions, be turned into a totally different kind of entity: heat, light, electricity, or a shock wave—all forms of energy? What was the perceived power of uranium before it was harnessed to destructive ends? Did the scientists intend to create a doomsday weapon, or were they simply pawns in an increasingly grotesque political theater at play? Could the horror of the atomic bomb have been averted?

What led to the birth of the atomic bomb was a sequence of unlikely scientific discoveries of the properties of the humble, gray element uranium which gathered momentum over a period of several decades and reached a fever pitch on the eve of war.

1

PHYSICS AND URANIUM

Uranium is the heaviest element found in our natural environment. With an atomic weight of 238 (or 235, for the rare form of this metal), it is in fact so heavy that it cannot be produced in the same way as light elements. Unlike many lighter elements, uranium is created in a supernova—a tremendous stellar explosion. Our solar system, including our own planet, was formed from the remnants of stars that lived and died in our cosmic neighborhood. The hydrogen and helium that were formed during the big bang burn inside stars through a nuclear process called fusion, which merges the nuclei of small elements to create larger ones. Thus carbon, nitrogen, oxygen, and all elements up to iron in the periodic table are produced inside stars. When a star with a mass the size of our Sun or even somewhat greater dies, it sheds its atmosphere, and the elements that had been produced by its nuclear flames dissipate into open space. Millions of years later, such resulting clouds of elements may condense, as happened when our solar system came into being about 4.5 billion years ago, and this is how many of the elements on Earth came to be. The clouds of matter from supernovae mix with those of the remnants of stars that died a less violent death, and this is how uranium found its way to our environment on Earth.

Uranium thus exists everywhere on our planet. It makes up a tiny percentage of rocks, as well as of sea water. But what *is* uranium?

The matter in our universe consists of atoms, which combine with other atoms to form the molecules of the substances we know from everyday life, such as water—two parts hydrogen to one part oxygen—or carbon dioxide—one carbon atom for every two atoms of oxygen. Each atom has a core, called the *nucleus*. The nucleus itself is much, much smaller than the atom as a whole. If an atom were the size of a bus, then the nucleus would be the dot on the letter "i" in a newspaper story read by a passenger on the bus.[1] The nucleus is dense and contains protons, which carry a positive electrical charge, and also electrically neutral components called neutrons.

Depending on the element, there will be a different number of protons and neutrons inside the nucleus. The rest of the atom is made up of electrically negative components called electrons, which orbit the nucleus. These orbits and the empty space they occupy form most of the volume of the atom.

Hydrogen is the simplest and lightest element in the universe: it has only one proton in its nucleus. Helium is larger and has two protons as well as two neutrons. Hydrogen has one electron in orbit around its nucleus; helium has two. Uranium is very heavy—it has 92 protons in its core together with 146 neutrons—and it has, as is usual for an atom, the same number of electrons as it has protons. Thus there are 92 electrons in orbit around the uranium nucleus.

What is unusual about the uranium atom is the very large number of neutrons, and they add significantly to the uranium atom's weight and make it prone to decay. Because it is so heavy and dense and because of conditions inside it, the uranium nucleus disintegrates slowly, producing radiation, mainly in the form of alpha particles, which are helium nuclei. In the process, uranium gives rise to other, lighter radioactive elements—which also disintegrate by emitting radiation—and eventually becomes (nonradioactive) lead. The typical, standard unit of time that it takes a radioactive element to disintegrate is called a half-life. This is the length of time it takes half the mass present to disintegrate through radiation—and uranium's half-life is very long. Half of any amount of uranium 238 will become lead after 4.47

billion years. The radiation from uranium produces heat energy in rocks deep inside the Earth, and this process helps keep our planet's core warm; uranium is thus responsible for some of our planet's geological activity.

Uranium forms natural compounds that have many beautiful colors: bright yellow, glowing orange, fluorescent green, dark red, and black. These shiny minerals caught the eye of ancient Roman artists, who used uranium compounds to decorate pottery and tint glass. Some attractive Roman glass urns containing uranium minerals for color have been found in Cape Posillipo, near Naples, during archeological excavations.

THE MODERN STORY of uranium began early in the sixteenth century when a major silver discovery was made in an area with thermal baths in the German principality of Saxony. The silver rush led to the founding of a town called St. Joachim's Valley, or Joachimsthal.

It soon became the largest mining center in Europe, with a population numbering 20,000; Prague, the nearest large city, had only 50,000 inhabitants at the time. Eventually, two million silver coins, called *Joachimsthaler* after the town, were minted for the Austro-Hungarian crown, which owned the mines. The Joachimsthaler, later shortened to *thaler*, became accepted in many countries and gave its name to our familiar monetary unit, the dollar.

In 1570, the emperor Maximilian II gave the order to exploit the Joachimsthal mines and find more silver and—he hoped—other valuable metals. Using an improved mining technology, within a few years bismuth and cobalt deposits were found. Then something strange was discovered. It didn't look like silver, or cobalt, or tin, or any other mined metal. It was a dark compound the miners named pitchblende, from the German words for black and mineral. Nobody understood its properties and it was ignored, pushed aside as mere tailings in the mining operation.

Martin Heinrich Klaproth (1743–1817) was trained as an apothecary, and spent most of his life working in pharmacies in various locations in Germany, finally settling in Berlin. Klaproth had a severe face and an exacting, punctilious nature. He was a successful businessman as well as a curious scientist. His ambition went far beyond mixing and dispensing medicines, and he began to study chemistry on his own. Klaproth devised new methods of analysis of chemical compounds, which led to the founding of the field of analytic chemistry. He had a special talent for treating minerals—dissolving them in hydrochloric and sulfuric acids, then oxidizing them or heating them— so that he could determine their composition. Within a few years of applying his methods, Klaproth was able to discover cerium (a rare-earth, silvery metal), and explain the composition of a number of compounds.

Klaproth heard a rumor that in Joachimsthal miners had found a strange new mineral, and this fired his interest. He traveled there to inspect the mysterious compound and took a sample of the material back to his shop in Berlin. He submitted the compound to various tests, attacking it with acids and oxidizing agents to uncover its nature. After months of hard and often frustrating work, in 1789 he managed to find the right mixture of chemical agents that allowed him to finally extract from the pitchblende something he described as "a strange kind of half metal." Inspecting the odd compound he had just created, he determined that it was an oxide of a metal that had never been seen before.

In 1781 the planet Uranus—named after a Greek god—had been discovered by the German-born English astronomer William Herschel. To honor Herschel's discovery, Klaproth named his new element *uranium*. It was a generous tribute since by scientific convention he could have bestowed his own name on the new element, which might then have been named klaprothium.

Klaproth's discovery of uranium and of other metals he isolated and identified established him as the greatest chemist in Germany, and one of the greatest of all time. In 1810 the University of Berlin created a chaired professorship for him.

Following Klaproth's discovery, uranium was identified in minerals mined in many places around the world, but the deposits known then were never as rich as those found at Joachimsthal. By the twenty-first century, regions in Canada, Australia, and the Congo would surpass the Saxony mine. By then, uranium also had been found in Cornwall in Britain, Morvan in France, and locations in Austria and Romania.

SINCE KLAPROTH HAD only synthesized an oxide of the new metal—uranium combined with oxygen—chemists wanted to see the actual metal in its pure state. They understood that what had been synthesized was a compound and not a pure element, as one would be able to tell the difference between a powder and a solid metal. They realized that the metal was very heavy and dense, but had difficulties in isolating it from its compounds found in nature. In 1841, the French chemist Eugène Péligot used a powerful thermal reaction, heating uranium oxide together with potassium to separate uranium from oxygen. He achieved this complicated task by first turning the uranium oxide into a salt, uranium chloride. Then he reduced the salt chemically using the potassium. As the potassium began to act on the uranium salt (because at the much higher temperature it was reactive with chlorine to a higher degree than uranium), Péligot suddenly saw a shiny metal appear. This was pure uranium. It looked like silver, but it quickly oxidized again in the air.

By the middle of the nineteenth century chemists knew definitively that a very heavy element, a metal, had been discovered. But what was its place among all the other known elements? How did it relate to other elements found in nature?

By the end of the eighteenth century, chemists had known how to distinguish two groups of substances: pure elements, such as the metal sodium, and chemical compounds, such as sodium chloride (common salt). But how should the elements be classified? Nobody had yet answered this question. Surely, there was some method to the elements' chemical reactivities—the way elements combined to form compounds. Then, throughout the nineteenth century, as chemistry advanced as a scientific discipline, many new chemical compounds and the pure elements synthesized from them were discovered with increasing frequency. But there was still a great disorder in our understanding of the elements—how they fit together in the universe, and how they relate to one another. Chemists were discovering some rules of behavior—which elements reacted with which—and were compiling a list of elements, which by 1830 numbered 55. Were these all the elements in the universe, or were there others? How many? Since the rules of behavior of the elements were not well understood, the list was not very meaningful. What was needed was a kind of table that could arrange all the elements in a logical way that reflected and demonstrated their reactions with one another.

The first steps toward a logical classification of the elements in chemists' lists were made starting in 1817 by a German chemist named Johann Wolfgang Döbereiner, who showed that when the elements' atomic weights were arranged in increasing order, there were elements whose weight fit in the middle between the weights of pairs of other elements. For example, strontium (weight about 88) fit between calcium (weight close to 40) and barium (weight about 137). He found a number of such triplets of elements and began to look for groupings of other chemical elements. Several chemists improved on that idea, but the real breakthrough was achieved by a visionary Russian chemist.

Dmitri Mendeleyev (1834–1907) was born in Siberia and became a professor at the University of St. Petersburg in 1865. In 1871, he completed his masterpiece: the periodic table of the elements. Mendeleyev arrived at the idea of the periodic table by trying to arrange all the known elements by their atomic weights and in a way that would somehow capture their shared chemical reactivities and their similar physical properties. His table classified all the known elements of the time by each one's chemical properties and increasing atomic weight, and it showed uranium with the highest weight of all the elements. The structure of the periodic table placed the elements naturally into groups that behaved in similar ways. Thus chlorine, fluorine, bromine, and iodine (called halogens) were placed in one column—they form similar chemical compounds (by taking or sharing a single electron, as we understand it today). Similarly, sodium, potassium, and lithium are metals that behave in common ways (they donate one electron each to form salts). Later it was discovered that it was the atomic number (the number of protons or electrons in an atom) rather than the weight (which incorporates the number of neutrons as well as the protons) that determines chemical activity. Even with these discoveries, the modifications to Mendeleyev's table were minimal.

Years later, laboratory-produced elements exceeding uranium in weight would be added to the table. These included plutonium, einsteinium, and mendelevium—the last two honoring Einstein and Mendeleyev.

Uranium held a privileged place in the periodic table. Literally, uranium is the element out of this world. It was created during the supernova explosion of a massive star, and it was the last element in the table, being the biggest and heaviest naturally occurring element. With atomic number 92, the valence (the number of electrons it shares or donates in chemical reactions) of uranium is 6 or 4. Thus, when purified in an industrial process to separate its different isotopes, it is made to react with six fluorine atoms to make uranium hexafluoride, which is a gas, and can thus be separated according to weight using a centrifuge. Pure uranium is a silvery white metal, and it is very heavy—it feels like a piece of lead, but it is not as dark and can be polished to a shine. Uranium is radioactive and decays into other elements. All of these elements, except for lead—the final outcome of some radioactive decay chains starting from uranium—are radioactive. But what is radiation? What is radioactivity? And how were they discovered?

No one was looking for radiation. Its discovery was one of the most serendipitous moments in scientific history and took place at dusk on November 8, 1895, in a laboratory at the University of Würzburg in Germany. Wilhelm Conrad Röntgen (1845–1923), a 50-year-old professor of physics, was carrying out a routine experiment with an electric tube he had invented, when he suddenly noticed that a chemically coated sheet of paper on a bench several feet away from him was glowing lightly. He was stunned. He turned off the electrical current in the tube, and the glow disappeared; he turned it on again, and it returned. Röntgen realized that he had chanced on a fascinating discovery—a glow that could be induced from afar. He surmised that unseen rays traveled from the tube to the paper, causing the glow. And, with more experimentation, he realized that the rays that produced the fluorescence were able to penetrate certain materials (paper, wood, and human flesh). Here was a technical application that made the wonders of the human body accessible. Before, you had to cut someone open to peer inside. Now Röntgen realized that with his newly discovered X-ray radiation, the insides of the body could be visible. This held great promise for beneficial use in medicine, and hence the great excitement about this amazing advance.

Röntgen spent many months studying radiation and discovered that lead shielding blocked the rays. He published his results on X rays (which in some countries are still called Röntgen rays in his honor) in a paper he read to the Würzburg Physical and Medical Society in December 1895 (translated and published in *Nature* in 1896). In 1901, he would receive the first Nobel Prize awarded in physics. Scientists the world over set out to investigate the new phenomenon. Two related questions that occupied many were: Does radiation occur naturally? Do any natural compounds give off similar radiation?

The French mathematician Jules Henri Poincaré (1854–1912) read the new scientific paper by Röntgen describing his discovery and experiments with X rays, and championed the findings at a meeting at the French Academy of Sciences in 1896. Eminent French scientists were fascinated by Röntgen's work. Among them was the physicist Antoine Henri Becquerel (1852–1908), who had been studying the way objects give off internal light or phosphorescence, such as the glow of a firefly or certain algae. Becquerel was then studying uranium salts in his lab. Poincaré suggested to him that if X rays could cause fluorescence, perhaps the glowing salts in his lab also emitted some kind of rays.

Becquerel took up Poincaré's suggestion and spent several weeks experimenting with the uranium salts. He could not detect any luminescence in the compounds. He wanted to take some photographs outside, but since the weather was inclement, he placed—by chance—his photographic plates in a drawer containing the uranium salts. A few days later, he took pictures with these plates and in developing them noticed something very odd: the plates were cloudy. Pondering this mystery, he concluded that the streaks had to have been caused by the uranium salts. Perhaps here was proof that the uranium salt was creating a radiation similar to X rays. (To this day, film is often used to detect radiation.) Becquerel presented his results—which by then he had confirmed using controlled experiments—to his colleagues at the Academy of Sciences, and in 1903 he would share a Nobel Prize with a married couple, who lived and worked across Paris from his lab, for their co-discovery of radioactivity.

It had then been established that Earth contained a strange element, uranium, which possessed the property of radioactivity: It emitted radiation that could be detected, but whose full nature was not understood. Scientists were set to uncover its mysteries.

2

ON THE TRAIL OF
THE NUCLEUS

Marie Curie (1867–1934) was born in Warsaw to parents who were both teachers. Keenly interested in science and mathematics, Marie had an intense ambition and driving urge to study and succeed. Women were not permitted to study in universities in Poland at the time. She worked as a governess and, with meager savings from her job, helped her elder sister to immigrate to France to study medicine at the Sorbonne. After her sister finished her degree and married, she, in turn, helped Marie to move to Paris. Marie resumed her studies at the Sorbonne at age 24 and later married a classmate, Pierre Curie, with whom she worked in a prominent French physicist's laboratory. In 1896, the Curies learned of Henri Becquerel's research and decided to study the new phenomenon of radioactivity. Becquerel was by then a famous scientist and revered by physicists in the Parisian scientific community. At that time, Marie was looking for a subject for her doctoral dissertation in physics while recovering from the birth of their first daughter, Irène. Pierre suggested that the subject for her thesis might be an accurate measurement of the mysterious rays emitted by Becquerel's uranium salts, which had not yet been studied quantitatively. She set to work and obtained uranium salts of the kind Becquerel had used. Her first results confirmed his finding that the intensity of the rays produced by the salts was proportional to the concentration of uranium in the compound.

Encouraged by the quick payoff from her first step in this research, she decided to investigate uranium ores rather than salts, theorizing that she could make greater progress by using the actual untreated compound. So she turned to Martin Klaproth's pitchblende from Joachimsthal. Working with a small amount of the pitchblende, Marie was surprised to find that the emission of rays from this ore was much more intense than she would have expected merely from its limited uranium content. This puzzled her: Something strange was causing a difference between the measurement from the raw ore and the measurement from the pure uranium salts. She tried to understand what was happening but experiment after experiment failed to provide a reason. She was frustrated and angry at her inability to find out why there was such a difference in the radiation levels.

Marie Curie was about to give up her quest and abandon radiation research altogether when she suddenly came to a surprising conclusion: The increased radioactivity was due to trace amounts of another, yet-unknown element inside the pitchblende ore that was far more radioactive than uranium. But now she had to prove this bold assertion.

Marie enlisted her husband to help her provide experimental evidence for what she believed was causing the greater radiation. Together, after many weeks of refining and extracting chemical compounds from the raw ore, they were able in 1898 to identify a minuscule amount of an entirely new radioactive element. The Curies named it polonium, after Marie's native country. Aided by a chemist, Gustave Bémont, the Curies continued their painstaking work and in the same year found within the ore a second new element that was even more radioactive than polonium. They named it radium. They realized that this new element was extremely rare, and that in order to extract any measurable quantities of it, they would need an entire ton of pitchblende. The Curies wrote to the Vienna Academy of Sciences, asking for recent residues left over from the extraction of uranium at Joachimsthal.

A few weeks later, a large horse-drawn cart arrived at the School of Physics and Chemistry of the University of Paris. Canvas bags filled with brownish residue mixed with rocks and tree branches were un-loaded, weighing a ton in total. This was only the first such transport of radioactive waste, and one wonders how, years later, the French were able to clean the center of the Left Bank from residual radioac-tivity. In fact, radiation damage was already evident. Marie's fingers were continually burned and never healed completely. Because every-thing having to do with radiation was still new and not fully under-stood, the Curies had no predecessors to warn them about the dangers of radiation. Marie worked with bare hands and took no precautions. She would eventually die of leukemia.

Arrangements had to be made to store the pitchblende under-ground in the basement of the lab. After a lot of work with the raw ore, using a powerful refining technique they had perfected, the Curies were able to extract a tiny but measurable amount of ra-dium—about a tenth of a gram. This was a major achievement, but what they now needed to know was the relationship between ura-nium, which constituted the major element in the totality of a full ton of pitchblende, and the minuscule amount of the newly discov-ered elements. The Curies were elated to solve an important mystery of science—the sources and the nature of radioactivity—and they published their results in April 1898; thereafter, they were seen as the leading experts in the world on radioactivity. Along with Henri Bec-querel, the Curies received the 1903 Nobel Prize in physics for their discovery of radioactivity.

THE SAME YEAR that the Curies identified traces of radium in Paris, the British physicist J. J. Thomson (1856–1940) discovered the electron—a negatively charged particle in the atom—when he noticed that a ray he produced in a cathode ray tube (a glass tube with a near-vacuum inside) was deflected in a magnetic field, thus indicating that the ray's constituent particles possessed negative electric charges. He was able to show that the atom contained positive elements (protons) and negative ones (electrons). The atom was no longer the smallest unit of matter. To further understand the atom's structure, research teams around the world undertook various experiments. The atom's neutral components (neutrons) would be discovered three decades later. In 1906, Thomson received the Nobel Prize for his discovery of the electron. The race was now on; scientists all over the world were trying to uncover further the secrets of the building blocks of matter. "What is matter made of?" was the question that consumed them.

Thomson had a bright student named Ernest Rutherford (1871–1937), a New Zealand–born physicist. After moving to Canada, Rutherford wanted to continue the exciting research of his former professor and learn more about the structure of the atom. His British-led group working at McGill University in Montreal attempted to determine whether polonium and radium, discovered in pitchblende by the Curies, were somehow linked to uranium by a process taking place within the uranium atom. Rutherford's work at McGill showed that the alpha rays produced by radioactive decay were made up of positively charged helium ions (the nuclei of helium atoms). Rutherford achieved this understanding by placing radioactive elements that emit alpha rays near vacuum tubes, which were empty. After a while, the contents of these tubes were analyzed and were found to include helium. The only way this could have happened was if the radiation entering the vacuum tube was made of positively charged helium ions, which—as they entered the empty tube—"captured" electrons from the glass surface and became helium atoms. For this important work furthering the understanding of the nature of alpha radiation, Rutherford was awarded a Nobel Prize in chemistry in 1908.[1]

Rutherford accepted a position at Britain's University of Manchester, whose physics laboratories rivaled J. J. Thomson's at Cambridge. He continued his work with renewed energy, devising experiments to probe further the nature of the atom. At that time, the atom was still thought to be a uniform ball of matter in which positive and negative particles were embedded together. He wanted to see what would happen if an atom was bombarded by the alpha particles whose existence he had discovered—specifically, would alpha particles be deflected back by the bombarded atoms? The idea behind his experiments was similar to playing billiards in the dark: You shoot the cue ball and see if it hits anything; if it hits a ball straight on, rather than sideways, sometimes it will bounce directly back to you. By directing a stream of alpha particles emitted by radioactive matter at a piece of material such as a sheet of metal, if some particles are detected coming back from the metal, then they must have hit something inside the metal's atoms to scatter them.

In Rutherford's experiments, about one in a thousand alpha particles that were "fired" at a thin sheet of metal recoiled back at an angle from their original direction of motion. This result implied that the makeup of the atom was not uniform throughout, but instead, as Rutherford reasoned, each atom contained a dense center surrounded by a lot of empty space. Rutherford named that center of the atom— which was the part of the atom that pushed back the alpha particles in the collisions—the atom's nucleus. Rutherford reasoned that the nucleus repelled the positively charged alpha particles because it also had a large positive charge (it was understood by this time that positive charges repelled positive charges and negative charges repelled negative ones).[2] Thus, the significance of Rutherford's finding was that the atom had an internal structure with a nucleus that was made of particles with a positive charge because they caused the positively charged alpha particles to bounce back.

Eight years later, in 1919, Rutherford conducted other ingenious experiments that demonstrated that in certain cases, when an alpha particle hit the nucleus of an atom, hydrogen would be discharged into a vacuum tube placed near the experimental apparatus. He reasoned that small positively charged particles were emitted, which later captured one electron each. This very basic, positively charged particle was the "opposite" of the electron, and Rutherford named it the proton (derived from the Greek word for "first one").[3]

The proton and the electron were understood to be the basic particles within the atom that were electrically charged, and their charges were understood to be opposite in kind: the electron is negative (by convention), and the proton is positive.

Scientists always look for ways of understanding the universe based on what they already know. Thus, the first model of the atom posited a system similar in its structure to what had been known since

Kepler's discoveries in the seventeenth century about the orbits of the planets around the sun, with electrons circling the nucleus. The indefatigable Rutherford noticed that the actual weights of atoms were often twice what he would have expected them to be based on the weights of the protons they contained. The proton is much more massive than the electron—one proton weighs as much as 1,836 electrons. The weights were deduced from the angle of deflection due to the effect of gravity on a ray of particles. In 1920, Rutherford hypothesized that the nucleus of most elements also contains another type of particle—one that must have a neutral charge. He coined the term "neutron" for such a particle, which he believed populated the centers of atoms, their nuclei, along with the positively charged protons.

A decade later, in 1932, James Chadwick proved his colleague Rutherford correct by discovering the neutron.[4] Based on the findings of Thomson, Rutherford, and Chadwick, scientists now understood that atoms contained dense and heavy centers, termed the nuclei, and that electrons orbited the nucleus of each atom, encompassing within their orbits the volume of the entire atom—which was much larger than that of the dense nucleus.

But this structure raised a conundrum: Why did the electrons not glide into the nucleus and combine with it? Clearly opposite charges attract each other, so what was there in the atom to keep the electrons from entering the nuclei? The answer to this question was given by the prominent Danish physicist and quantum theory pioneer Niels Bohr (1885–1962), who constructed a more sophisticated model of the atom based on the emerging principles of quantum theory.

NIELS BOHR MADE enormous contributions to physics research and the basic ideas of quantum mechanics. He was born in Copenhagen; his father, Christian Bohr, was a well-known professor of physiology, and his mother was Ellen (Adler) Bohr, the daughter of the financier D. B. Adler, founder of the Commercial Bank of Copenhagen. The couple enjoyed the company of intellectuals and invited many thinkers and scientists to their home. Their two boys, Niels and

Harald (who became a prominent mathematician), grew up absorbing ideas through stimulating conversations. Niels chose physics and became a leading figure in the field, the man around whom an entire community of physicists developed—maturing in their scientific thinking and producing some of the most important research results. As a result of this prominence, Bohr was chosen to head the Institute for Theoretical Physics of the University of Copenhagen, which was financed in large part by the Carlsberg Foundation in Copenhagen.

In 1913, Bohr constructed an atomic model based on a hypothesis put forward around 1900 by Max Planck that blackbody radiation (the energy emitted by a glowing body) was absorbed and emitted by atoms in specific, discrete packets that Planck called quanta. His assumption immediately explained the experimental results about such radiation. Bohr hypothesized, analogously, that the electrons in an atom orbited the nucleus only in orbits with precisely determined, "quantized" levels of energy.

Thus, the electron travels around a nucleus in an orbit with a given radius (and hence a given energy level) and can drop down only to another, specific orbit with a lower energy level—and not to an arbitrary orbit that might exist between them. In this way Bohr did away with the idea of any continuity of energy levels for an electron, in the same way that Planck had eliminated a continuum of energy levels for blackbody radiation. In both situations, only specific levels of energy (quantized levels) were possible.

This quantization of orbits prevents the electrons from sliding down to the nucleus. When an electron jumps down from one quantized energy level to another, it emits the energy of the difference between the two levels (the two possible orbits) in the form of a particle of light, a photon. Bohr had thus used the quantum theory to account for observed phenomena—the discrete energy levels of light—and to greatly improve the understanding of the structure of the atom. But much work lay ahead in the first decades of the twentieth century, and scientists were eager to understand the mystery of the atom, of radioactivity, of energy, and of mass. And they wanted to know where matter came from and where it was going.

ERNEST RUTHERFORD HAD been the first to explain that radioactivity is produced by the decay of atoms. In 1902–1903 Rutherford and Frederick Soddy had analyzed the decay products of various radioactive elements and determined that radioactivity was caused by the breakdown of the atoms in the element, and produced a new element. (Soddy would continue this work and would later be credited with the discovery and clarification of isotopes.) With another member of his scientific team, Bertram Boltwood, in 1904 Rutherford had worked out the transformation that radioactive elements undergo, estimating the rates of change involved. Their sophisticated analyses of the radiation obtained from uranium had shown that uranium has relatively weak radioactivity, decaying very slowly, at the rate of one milligram per ton per year.

Uranium is transformed into inactive lead through a chain of radioactive elements, each of which has a characteristic disintegration rate. Uranium 238 becomes thorium 234, which further disintegrates by radiation to form radium; then the radium disintegrates to become the gas radon, which in turn gives rise to polonium. Polonium is the last radioactive element before lead is produced. Since lead is not radioactive, the process of radioactive disintegration stops with it.

For every three tons of uranium refined from ore, only one quarter of a milligram of polonium and one gram of radium are obtained. Rutherford's discovery thus provided further support for what the Curies had found in their lab: It explained the transition from uranium to lead through thorium, radium, radon, and polonium. Exactly what was happening inside the uranium nucleus remained the mystery to be solved. The key work on this question would be done in Paris.

THE CURIES HAD meanwhile continued their experimentation with vigor. In 1904 they completed the separation of one gram of radium from a further eight tons of Joachimsthal residues. Around this time medical applications of radium were being discovered, and Pierre Curie collaborated with French physicians in a study that resulted in widespread use of radiation in treating cancer. The "Curie therapy" entailed implanting radium into tumors to shrink them.

The Curies had become the celebrity couple of France. Marie was the first woman to obtain a Ph.D. in science in France and the first to win a Nobel Prize. But the important work on radioactivity pioneered by the Curies was interrupted in 1906 when Pierre died in a traffic accident. Marie was subsequently appointed to her late husband's professorship at the Sorbonne, becoming that institution's first female teacher. In 1910 her fundamental treatise on radioactivity was published. And in 1911 Marie Curie was awarded a second Nobel Prize, this time in chemistry, for her discoveries of polonium and radium and her work isolating the pure form of radium.

By then, the buzz in the scientific community about radiation and its effects had reached a fever pitch. For the first time, scores of scientists in many countries were working on understanding radioactivity and the structure of the atom. These scientists' experiments were intertwined in complicated ways that transcended national borders more than in any previous period. Uranium and its strange properties had focused the attentions of a group of brilliant minds on the quest to uncover its secrets.

SCIENTISTS WERE SEEING things that defied the laws of classical physics. Since the discovery of the atom, it had been assumed that atoms in nature were immutable and unchangeable. But radioactivity hinted that this could be wrong. That uranium released radiation and in the process transformed itself—first into radium, then polonium, then lead—indicated that matter was mutable. The medieval alchemists might have felt somewhat vindicated—but instead of turning elements into gold, modern alchemy was turning uranium into lead. However, the actual process of uranium's radiation and transformation into other elements presented several scientific puzzles. How was it that uranium disintegrated into other elements, and why? What was causing an atom to change into other atoms? There was great hope in the world's scientific community that a major meeting might help point the way to solutions.

Ernest Solvay (1838–1922), a Belgian industrialist who had made a large fortune in the chemical industry, had a passion for science. In 1911 he organized and financed an international conference held at the Metropole Hotel in Brussels (the only nineteenth-century hotel still standing in Brussels). This Solvay Conference hosted twenty-one of the world's most prominent scientists to discuss the study of radiation and the structure of the atom. Among those present were Henri Poincaré, Ernest Rutherford, Marie Curie, Paul Langevin, Max Planck, and, the youngest member of the group, the 32-year-old Albert Einstein. The meetings led to important collaborations in the budding area of quantum theory—which would be developed more fully starting in 1925 by the Austrian physicist Erwin Schrödinger, the German mathematician and physicist Werner Heisenberg, the English physicist Paul A. M. Dirac, and by the Austrian-born physicist Wolfgang Pauli.

Ultimately, the model of the atom we have today is based on quantum theory. Quantum mechanics posits that the location and velocity of atomic particles are impossible to define together. All we have are sets of probabilities for such parameters. It is not always possible to attribute cause and effect, and this runs against our conception of logic in nature.

The quantum mechanics picture of the atom, developed in the 1920s and later, differs from the traditional Bohr model of the atom as a mini solar system. The electrons, unlike planets, do not have well-defined positions in their orbits at any given moment in time. They follow quantum rules that allow us only probabilistic knowledge; the uncertainty principle, proposed by Werner Heisenberg in 1927, prevents us from ever knowing everything at once (position and momentum, or location and speed, or time and energy) with perfect precision. And the rule in quantum mechanics is that a particle can be here *and* there at the same time, not necessarily here *or* there. The quantum picture of reality is fuzzy and intrinsically given to the laws of uncertainty and probability.

Einstein's equation $E=mc^2$ was recognized by the scientists at Solvay to be a key tool for approaching the riddles of radioactivity, radiation, and atomic disintegration. According to Einstein's famous formula, the atom, because of its mass, contained much energy. Atomic disintegration was one manifestation of Einstein's mass-energy equivalence, as evidenced by the release of energy from the uranium mass through radiation. But what was the process? If it could be understood, then radiation and atomic processes might be developed to produce considerable amounts of energy. Surprisingly, this important possibility about the nature of the atom—which would eventually lead to nuclear bombs and peaceful atomic power generation—came as early as 1911 at Solvay. With Einstein's theory, uranium would now provide clues to our understanding of energy.

Out of every 1,000 atoms of uranium found in nature, 993 are a type called uranium 238, and only seven atoms are of the rare type called uranium 235. These two isotopes of uranium behave the same chemically, but they are different in their nuclear properties. Both kinds of uranium have 92 protons and 92 electrons, making their chemical reactivity identical. But uranium 238 has 146 neutrons in its nucleus (along with the 92 protons packed together with them), while uranium 235 has only 143. Thus, the atomic weight of uranium 235 is smaller than that of uranium 238 by 3 units of mass (the weight of 3 neutrons, which is slightly more than the weight of 3 protons, but we ignore this very small difference).

These two isotopes of uranium are also different in their stability. Uranium 235 is far more unstable than uranium 238. If we have equal quantities of uranium 235 and uranium 238, at any given moment uranium 235 will have more than six times as many atoms disintegrating by radioactive decay than does uranium 238. This property is reflected in the half-life of each of the two isotopes.

The term "half-life" was invented in 1904 by Ernest Rutherford. As noted earlier, the rate of radioactive decay is calculated as a percentage: in a given unit of time, a certain constant percentage of the nuclei of any particular radioactive element decay through radiation. Therefore, the amount of original material that is undergoing radioactive decay is declining exponentially. To understand what happens with radioactive decay, think of a bank account containing $100, and suppose that the bank charges you a fee of 1% per month just for holding the money. After one month, you will have $99. After two months, the bank will have taken another 1% of what is left over from the first month, or $99 × (0.01) = $0.99, and you will be left with $98.01. A month hence, you'll be left with $97.03 and so on: your initial amount of $100 is declining exponentially at a rate of 1% a month.

This is exactly the same kind of process that characterizes radioactive decay. The question that Rutherford asked was: How long does it take for exactly *half* the original amount of matter to decay? Or, in this money analogy, how many months does it take for you to be left with only (or as close as possible to, since this is a discrete, once-a-month deduction process) $50? In this example, the answer is 69 months, or 5 years and 9 months. Thus, the half-life of your account is 69 months.[5]

This half-life measure can be used to compare radioactive elements based on their rate of decay. The rate of decay is also generally related to the intensity of the radiation emitted, and it is a measure of the instability of the nucleus. Elements with short half-lives are more unstable than those with long half-lives, and they radiate with greater intensity (because their nuclei decay faster—emitting more radiation for a given number of atoms in a given unit of time).

The half-life of uranium 238 is about 4.5 *billion* years. This isotope of uranium decays slowly. Uranium 235, by contrast, has a half-life of only 700 *million* years. Thorium 232 is an element that decays even more slowly than uranium 238. Its half-life is 14 billion years. In the entire time span since the Earth was formed 4.5 billion years ago, only one out of six thorium 232 atoms has decayed.[6]

But many elements have much shorter half-lives than even uranium 235, and they decay in terms of hundreds of years, rather than millions of years; others have half-lives of only a single year; and then there are elements with half-lives of months, of days, and some even of seconds. The half-life of radium 226, an isotope of the radium discovered by the Curies, is 1,620 years. Since the frequency of change is quick compared to that of the Earth, scientists realized that radium had to be constantly created or by now there would be none to find. This replenishment happens through the decay of uranium.

Polonium has an even shorter half-life. There are a number of isotopes of polonium as well, but the longest living of them all is polonium 209, whose half-life is a hundred years. And francium, an element discovered by the French physicist Marguerite Perey and named after her native land, has an even shorter life. Its longest-lived isotope, francium 223, has a half-life of only 21 minutes.[7]

AT THE 1911 Solvay Conference, Marie Curie, because of her prominence in radioactivity research, led many of the discussions. But while she was in Brussels, a scandal was brewing in Paris. A French newspaper printed love letters it claimed had been exchanged between Marie Curie and her research colleague Paul Langevin (1872–1946). Public outrage over this alleged affair followed. Segments of the French press accused her of being a foreign, Jewish (she was not) home-wrecker. Their anti-Semitic accusations echoed those of the Dreyfus Affair (in which a Jewish military officer was wrongly sentenced to imprisonment for spying) of a decade and a half earlier. A smear campaign had thwarted her nomination to the French academy of sciences earlier in 1911 and this one nearly jeopardized her second Nobel nomination.

On November 4, 1911, just as Curie and Langevin were returning to Paris, the French newspaper *Le Journal* wrote: "The fires of radium which beam so mysteriously have just lit a fire in the heart of one of the scientists who studies their action so devotedly; and the wife and children of this scientist are in tears." Marie Curie and Langevin separately tried to disappear from the public eye until the scandal blew over.

The Langevin affair caused a shift in the research activities at the Curies' lab and the First World War further interrupted their work as so many scientists were diverted to war-related activities. Curie herself worked extensively to equip mobile field hospitals and over 200 locations with X-ray equipment. She used the money from her second Nobel award to fund other researchers and the Radium Institute at the

University of Paris, where she returned to work after the war. Her daughter Irène, who would eventually make her own great discoveries about radioactivity, joined the laboratory in 1918. From 1922 on, Marie concentrated her research in the area of the chemistry of radioactive substances and the medical applications of radiation.

The research conducted by Marie Curie and her daughter required radium, which came from uranium disintegration. But radium, because of its extremely low concentration in pitchblende and the many chemical steps necessary for its extraction, had become the most precious substance in the world. Its price reached 750,000 golden francs (about $10 million in today's values) per gram.

Entrepreneurs caught onto the market potential of uranium ore. It was originally produced from the Bohemian ores in two French factories, which enjoyed a short-lived monopoly. But then the Austro-Hungarian government forbade the export of the ore and built a radium factory in Joachimsthal, next to the one producing uranium color compounds. The Austrians aimed to create their own monopoly, but because of the existence of mines in Britain, France, and Portugal, they failed. In 1913, the United States entered the global uranium market. Seeking radium for medical applications and research, American producers opened new mines in metal-rich Colorado. The actual extraction of radium from uranium ore was done in Pennsylvania by the Standard Chemical Company of Pittsburgh, which over 13 years, through 1926, put on the market a total of 200 grams of radium and 600 tons of uranium. About half the radium went to hospitals and the rest was used in luminous paint for watch dials.

U.S. production retained the greatest market share of radium and uranium for almost ten years. Then the Belgians made a discovery in Africa that brought about major changes. In 1915, a prospector at Shinkolobwe in the Belgian Congo discovered a deposit of pitchblende and other uranium minerals of a higher grade than had ever been found, and higher than any found since. Its discovery was kept secret by the Belgian Union Minière du Haut Katanga, which mined rich resources of copper and cobalt in the region. After World War I a factory was built at Olen, near Antwerp, and seven years later the secrecy was lifted with the announcement of the production of the plant's first gram of radium. The Belgian output of radium was as large as its cost was low, which convinced Standard Chemical to drop out. Union Minière then enjoyed a near monopoly and could dictate the price. Congo today remains a major producer of uranium ore.

Despite this increased availability of the minerals, the Curies and other European scientists had difficulty procuring it. The French government was not very helpful in funding Marie Curie's work, and she had to devote much of her time to raising money for her institute. The Curies and their co-workers nonetheless had made tremendous advances in our understanding of the phenomenon of radioactivity, had discovered new radioactive elements, and had made important contributions to medicine by providing the necessary understanding that radium, because of its radioactivity, could be used in treating cancer by destroying malignant tissue cells. It was now left for other researchers to continue this work exploring the mysteries of radiation. A young Austrian woman would play a key role in further expanding our knowledge of radioactivity.

3

LISE MEITNER

Three years before Wilhelm Röntgen's groundbreaking discovery of X rays in 1895, a gifted and ambitious 14-year-old girl received her graduation papers from a girls' school in Vienna. She would struggle to enter the male world of physics and would become a researcher and theorist. Lise Meitner (1878–1968) was born into a secular Jewish family in the city's Second District, the former ghetto whose walls had been brought down by the enlightened emperor Josef II some years earlier. Lise loved music and showed unusual promise in mathematics and science, but in turn-of-the-century Austria, girls left school at 14 and were expected to marry and keep a home.

Lise was eager to follow a different path. She was passionate about physics, but her parents wouldn't hear of further study in the field, as women were not allowed to study physics or any science in the Austro-Hungarian empire, which had only a handful of professional physicists. Grievously disappointed, she dutifully followed her lawyer-father's advice and enrolled in a three-year program that would result in a teaching certificate in French. Thus began a period she would later call her "nine lost years."

Lise was petite and very slender, with serious dark eyes that often had a faraway expression. Despite her youthful appearance she was more mature than, and stood out from, her peer group by her interest in the sciences, devouring books on chemistry, physics, and mathematics in her spare time.

When Röntgen's 1895 discovery rocked the scientific community and Becquerel's finding of uranium radiation followed in 1896, eighteen-year-old Lise Meitner became determined that this was to be her path in life: She would study radioactivity. Nothing could stop her, she resolved, not her parents and not a society that didn't allow women at the university. By the time she turned 21, Austrian laws had begun to change, and women were allowed into university programs. To make up for lost time, her father hired a tutor, and over two years of intensive work, the young Meitner was able to cover the courses boys took at a gymnasium, an academic high school. In 1901, a few months before her twenty-third birthday, the ambitious young woman—five years older than the boys entering her class—enrolled at the University of Vienna to begin preparing for what would become a brilliant career in physics.

Partly because she was older than the other students and partly because she was a woman trying to compete in a scientific world that was vastly dominated by men, she developed a stubborn drive to succeed. She studied all the time and spent as much as twenty-five hours a week in lectures and labs.[1]

Meitner took physics, chemistry, botany, and mathematics courses, working herself into exhaustion. Fortunately, her calculus class was early in the morning and she was fully awake and paying close attention to the lectures. Her calculus professor understood that he had an excellent student and assigned Meitner special tasks, which he thought would challenge her and encourage her to go beyond the standard material he covered in class.

To test her abilities, one day the professor asked Meitner to read an article by an Italian mathematician and find an error in it. He then suggested that Meitner publish her findings. But she refused. She was uncomfortable with the idea of publishing something she did not believe was fully hers, since her professor had obviously known that the error was there in the first place, and had set her up to find it; surely, the credit should go to him. This incident highlights her moral and ethical nature; throughout her life, Lisa Meitner would follow strict self-imposed codes of professional behavior. When she felt that a scientific result was even partly shared with someone else, she would insist on that person getting equal credit. Unfortunately, people she worked with later in her career would sometimes fail to reciprocate the professional courtesy they received from her.

Meitner's physics course was taught by Franz Exner, who was a friend of Wilhelm Röntgen and had a great interest in radiation and radioactivity. He had helped the Curies obtain pitchblende from Joachimsthal and received from them small amounts of radium in exchange. He was responsible for making Vienna a center for research on radioactivity in the first few decades of the twentieth century. The introductory physics course and its enthusiastic professor further convinced Lise that she wanted to be a physicist.

In her second year, Lise Meitner came under the influence of an even greater physicist—Ludwig Boltzmann (1844–1906), the famous pioneer of statistical mechanics, thermodynamics, kinetic theory, and atomic theory. Lise found Boltzmann's lectures the most beautiful and exciting of all, and she arranged her schedule so she could take all her physics courses from him.

Boltzmann was favorably inclined toward women in the sciences, recognizing the intensity of the discrimination against them. Some years earlier he had married a young woman, Henriette von Aigentler, and helped her fight similar prejudice when she had been denied permission to even audit courses at a university.[2] Boltzmann had helped her appeal that unfair decision. He therefore welcomed the ambitious Lise Meitner, whom he understood to have had little pre-university training. Eventually, Meitner met Boltzmann's wife, and they were able to share their experiences of struggling against the odds to study physics.

Baltzman's statistical mechanics, a branch of theoretical physics, relies on the proposition that statistical methods can explain the collective behavior of large collections of unseen atoms. In a scientific paper published in 1870, Boltzmann had shown how the application of the mathematical theory of probability along with the laws of mechanics could explain the exchange of energy stated in the second law of thermodynamics. Further, he derived an equation for the change in the distribution of energy among atoms in collisions. More broadly, he concluded that phenomena that were states of change or fluctuation could be understood using probabilities and the collective properties of the constituent unseen atoms.

Boltzmann's critics, however, refused to believe in any phenomenon or entity that could not be seen. Throughout his life, he had to defend his theories, and perhaps because the physics establishment before the turn of the century rejected his pioneering ideas, he was given to periods of depression. But his students uniformly adored him. He was a charismatic teacher who made his listeners question conventional beliefs and open their minds to the wondrous world of physics.

The discovery of radioactivity and of the existence of particles much smaller than atoms boosted the acceptance of atomic theory, and thus early twentieth-century discoveries gave weight to Boltzmann's conclusions. The deep implications of his groundbreaking work in physics, as well as his philosophy of nature, would continue to guide Lise Meitner in her career and exert the strongest influence on her thinking.

In 1905, after finishing her coursework at the University of Vienna, Lise Meitner embarked on her doctoral dissertation project. Because Boltzmann was at that time in California, she chose to work with her first physics professor, Franz Exner. She developed an experimental, rather than theoretical, project mostly because her previous studies had been confined to theory only, and she now wanted to experience laboratory work in order to round out her background and preparation.

In her doctoral work, Meitner discovered that Maxwell's laws of electromagnetism could also apply to heat conduction. Her dissertation, titled "Conduction of Heat in Inhomogeneous Solids," reported theoretical results derived from her experimental work, in which droplets of mercury were suspended in a medium made of fat molecules, and heat conduction through this substance was measured by a series of thermometers. This complex research project trained Meitner in the use of very involved experimental designs, which would become the hallmark of her future work on radioactivity. In December 1905 Meitner defended her dissertation, and in 1906 she became the second woman to obtain a Ph.D. at the University of Vienna.[3]

Meitner now faced the predicament of any woman in science in those days: the immense odds against getting a job in her field. She knew about Marie Curie's work in Paris, and so she wrote to her to inquire about the possibility of a position in Curie's lab. Unfortunately, there was no such opening, and similar disappointments followed. In order not to have to leave physics altogether, Meitner took a job for which she was overqualified, that of instructor at a girls school. But at night she continued her research work in the university's physics laboratories.

During 1906, Meitner turned her attention to research on radioactivity. The highly publicized achievements of the Curies, and a course she had taken on radioactivity, served to stimulate her interest in this area. She began to conduct experiments in radiation at Boltzmann's physics institute. Her first experiments consisted of measuring the absorption levels of both alpha and beta particles by thin foils of different kinds of metals.

In the fall of that year, the physics institute and the entire university were shocked when Boltzmann committed suicide at age 62. The physics community tried to come to grips with this tragedy: Some believed that Boltzmann's mental health had suffered because of the immense resistance to his ideas within the conservative older generation of physicists, others, that he was simply ill. Lise Meitner was especially affected by the loss, since she saw in Boltzmann a close mentor, supporter, and someone who understood her ambition to become a physicist despite the prejudice against women. His passing strengthened her resolve to continue to struggle for what she believed.[4]

Boltzmann's institute at the University of Vienna was directed for a time by a young physicist, Stefan Meyer, who was also interested in radioactivity and had conducted some experiments with Meitner. Meyer had investigated the properties of the first new radioactive elements discovered by the Curies in Paris and by the French chemist Debierne: polonium, radium, and actinium. He had actually discovered that beta radiation consisted of negatively charged particles, a finding that has also been attributed to the French scientist Henri Becquerel and the German physicist Friedrich Giesel. Becquerel later was able to show that these particles were electrons.

Lise Meitner then resumed her earlier work with Meyer on radioactivity. The number of known radioactive elements was continuously rising, and there was great interest in the mystery of radiation throughout the scientific world. What was radiation? What made an element radioactive? What was the nature of radioactivity, and how did it affect other elements? These were important questions, because they appeared to hold the key to the mystery of the constitution of all matter in the universe. The only equipment one needed to have in order to investigate radiation at that time was a device that measured the level of radioactivity.

Hans Geiger's radioactivity counter (the "Geiger Counter") was still some years away, and physicists in Vienna and elsewhere were using a more primitive device, the electroscope, to detect and measure radiation. It consisted of a very thin gold leaf attached at an angle to a metal rod, both placed inside a sealed glass tube containing gas. The rod extended to the outside of the tube, and could thus receive an electric charge from an external source. The charge made the gold leaf move away from the section of rod inside the tube, since like charges repel each other. But when a radioactive source was placed close to the tube, the radiation (positively charged alpha particles or negatively charged beta particles) ionized the gas inside the tube. This affected the gold leaf—because its own electric charge was now changed by its contact with the ionized gas—and it would now bend either more, or less, than its original angle, depending on whether the ionized gas was positive or negative. A scale attached to the device allowed for a measurement of the change in the angle of bending of the gold leaf, and thus provided a crude estimate of the intensity of the radiation.

In the fall of 1906, Lise Meitner used the electroscope in investigations of the absorption of alpha particles by sheets of various metals—a study she had begun earlier. A number of physicists were interested in this problem and wanted to know whether alpha particles were only absorbed by the bombarded metals or whether some of the particles bounced back. As noted earlier, both Ernest Rutherford and Marie Curie had reported evidence of the scattering of alpha particles, but these findings were disputed by other scientists who thought that none of the alpha rays ever returned—that there was no scattering, only absorption.

This question was important because its answer could shed light on both the structure of the alpha particles themselves and on the makeup of the atoms in the target metal foils at which the particles were aimed. Meitner's experimental results were in agreement with Curie's and Rutherford's. She confirmed that some alpha particles were indeed scattered. In fact, she even found that the rate of scattering was proportional to the atomic mass of the metals that were bombarded. This meant that heavier metals—the ones with more massive nuclei—were causing more scattering of the radiation particles. She published this result in the leading German physics journal, *Physikalische Zeitschrift*, in June 1907.

Following this successful experimental work, Meitner still felt that, other than teaching school and doing some lab work at the university after hours, she did not have a future as a physicist in Vienna. The city offered few prospects for an ambitious young physicist who had already made contributions to the area of radioactivity research. And she was unhappy to still be dependent on her parents for support since her income was not even sufficient for her modest living requirements. In late summer 1907, she asked her parents for money to travel to Berlin where, she knew, physical science was a major area of study at the university. She hoped to spend some time there learning new methods and absorbing new ideas. Lise was grateful that her parents agreed to give her additional cash for the trip, and she left for Berlin that September.[5]

Berlin in 1907 was one of the world's main centers for the exact sciences, but here too academia was dominated by men. Lise Meitner asked the greatest physicist at the Friedrich Wilhelm University, the discoverer of the quantum, Max Planck, if she could listen to his lectures.

Planck was kind and polite, and invited her to his home. Then he asked her why, since she already possessed a doctorate, did she want to attend his classes. Meitner responded that she sought to improve her knowledge of physics; and he summarily dropped the subject. She naturally assumed, therefore, that he was not interested in helping her at all. At that time, there were different views among professors at universities about the potential for women to do well in the sciences: Some believed that women could succeed, while others clung to the traditional prejudice that science was for men only. Planck's views probably fell somewhere between these extremes. He thought that very gifted women should be allowed in science, but not those with lesser abilities. Although Meitner did not understand it at that time, Planck did recognize her excellence, as evident by her doctoral degree and her preparation and publications, and in fact he was pleased to have her attend his lectures.[6]

Toward the end of September, Meitner met Otto Hahn (1879–1968), and they found that they shared a great interest in radioactivity research. Hahn was a chemist and lacked the physics and mathematics background needed for a full understanding of radioactivity, while Meitner's preparation was in physics and mathematics and only somewhat in chemistry. So their skills were complementary, allowing for the potential for good joint work at the intersection of physics and chemistry: the study of the behavior of radioactive elements. Hahn was working in the chemistry department of the university, and he introduced Meitner to the director of his department, who generously offered her the opportunity to work there as well.

Granted a research position at the Berlin university, Meitner now had to face both a sexist and a xenophobic academic atmosphere in Germany. Meitner's unprecedented appointment—an Austrian Jewish woman awarded a research position in a leading university—engendered great hostility. She was therefore excluded from the main laboratories of this institution, and was given instead a converted carpenter's bench in the building's basement as her work station. Meitner was also barred from attending any of the lectures presented upstairs.

But despite these shameful and degrading employment conditions, Meitner thrived. She had an ever-positive outlook on life, and a gift for forming new friendships and keeping friends for life. Hardworking and personable, she captivated her close colleagues, and in a few years, her admirers included Niels Bohr, who then tried actively to recruit her for his thriving laboratory in Copenhagen.

But Meitner chose to remain in Berlin. At the university, she had teamed up with Otto Hahn, and the two scientists were deeply involved in a major research project—one that promised to yield key results in the study of the nature of radioactivity and the structure of the atom. Their aim was nothing less than to make plain the secrets of radiation and explain internal changes in elements as a result of external radiation.

Theirs was a striking professional alliance of a handsome, patriotic young German man and an attractive and brilliant foreign Jewish woman; it was also a fortuitous wedding of physics and chemistry. To work with Meitner, Hahn had to descend from his capacious office upstairs to the basement so the two of them could conduct experiments as a team.

The two scientists worked many long days together. Despite their evident mutual attraction, they never became intimate. When they lunched together, it was always in the lab. The pair published prodigiously and established a body of work on radiation—the next generation of widely accepted results after those obtained by the Curies. Their papers all bore both their full names.

In December 1908, after Meitner and Hahn had been working together for some time, Ernest Rutherford passed through Berlin on his return trip from Sweden, where he had just received his Nobel Prize, to Manchester, England. He was introduced to the two young scientists now continuing his own pioneering work on uranium. But when Rutherford shook Meitner's hand (whose full name he had certainly seen on published papers), he said to her: "Oh, I thought you were a man!" He then turned away and spent his time with Hahn, leaving Meitner to accompany Mrs. Rutherford on shopping forays in Berlin.

4

THE MEITNER-
HAHN DISCOVERY

O n October 23, 1912, the Prussian government inaugurated the
Kaiser Wilhelm Institute for Chemistry in a suburb of Berlin. Here
chemistry was the main field of research, and for decades its top
practitioners were German. Both experimental and theoretical physics
lagged far behind in prestige in Germany. Although World War I would
demonstrate the utility of Röntgen's X rays to reveal bone fractures and to
find bullets lodged in the bodies of wounded soldiers, theoretical physics
would still be deemed a lowly discipline.

The new institute was modeled after Oxford, in a tree-lined neighborhood with parks where scientists and scholars could stroll and discuss ideas.
Otto Hahn became a senior member of the institute and had a substantial
salary. This enabled him to consider marriage and starting a family, and he
wed a young woman he had met in 1911, Edith Junghans.[1]

Lise Meitner's fate changed completely once the new institute was inaugurated. She became a key associate in the laboratory she shared with
Hahn. It was even named after them: the Hahn-Meitner Laboratorium. She
was bumped up on the pay scale, although her salary stayed behind Hahn's.
Their finances improved beyond the pay they were given for their work at
the institute due to additional income. A few years earlier, Hahn had discovered a highly radioactive element, an intermediate radioactive isotope of

thorium that he named mesothorium. The intensity of gamma rays emanating from this new source made mesothorium a good substitute for radium in medicine. From 1913 through 1914, Hahn was paid more than 100,000 marks for a quantity of this element purified in his lab, and he gave about a tenth of that income to Meitner, who had collaborated with him on this project.[2]

The Hahn-Meitner Laboratorium consisted of four large rooms on the ground floor of the Kaiser Wilhelm Institute of Chemistry's north wing. The two scientists kept their laboratory exceptionally clean although the health hazards of radioactivity were not well understood by science at that time.

On any given day in laboratories across Europe in the second decade of the twentieth century, many physicists were at work trying to understand radioactivity. Experiments with easily detected radiation by this time had been completed. Much of the research now focused on why some elements give off energy and what the various radioactive elements that physicists had discovered and named had in common. It would be several decades before all the strands of the mystery would lead researchers back to uranium. For now, clues to the bigger mystery were popping up piecemeal around Europe.

In 1912 Hahn and Meitner embarked on an ambitious research effort. In 1899, the French chemist André-Louis Debierne had isolated from pitchblende a new radioactive element he named actinium, from the Greek word *actinos,* meaning ray. This element (atomic number 89, weight 227), similar to titanium and thorium in its chemical activity, is extremely radioactive—it is 150 times as powerful a radiation source as radium. It emits both alpha and beta particles, meaning that its radioactivity consists of helium nuclei and electrons, and its half-life is about 22 years. Hahn and Meitner wanted to know where actinium came from. The first question scientists confronted in the lab when studying an element was whether the element was naturally occurring or a byproduct of another element. Since actinium was synthesized from pitchblende, Hahn and Meitner reasoned that it derived from uranium. Yet physical principles told the two scientists that it could not have come directly from uranium—there must have been some intermediary. The pair was searching for the "mother element" of actinium. This problem turned out to be extremely complicated and very hard to solve.

What motivated scientists to undertake such complex research projects that could take many years to complete, for which no solution was guaranteed, and for which the effort could be immense? Hahn and Meitner, along with others doing similar work in labs and institutions in Europe and America, were pioneers. Mendeleyev had laid out a map of elements more than forty years earlier, yet he knew the periodic table was not complete. He had reasoned that elements could be ranked according to the ascending number of atomic weight. He had left room in his scheme for expansion, assuming that elements must exist that would have atomic weights between those that were then known. The chart screamed out for completion. Finding a new element would be a plum prize for any researcher, and the challenge of discovering one, combined with the lure of fame, was enough to motivate a chemist or physicist to spend much time on such a quest. The scientists of the first few decades of the twentieth century, much like many scientists today, were obsessed with one goal: to uncover the stuff of the universe, to understand the mechanisms behind all that surrounds us.

THE MYSTERY OF radiation was a powerful key in the quest to solve the puzzle of the creation of matter. Radioactive elements are, by definition, *unstable* elements. Their nuclei undergo radioactive decay, in which a part of the nucleus is shed in the form of radiation: alpha rays (or particles—actually helium nuclei, two protons and two neutrons held together), beta rays (or particles, usually electrons), and gamma rays (high-energy photons, or particles of light, beyond the visible spectrum and thus with higher energies).

Since the radioactive elements are unstable, they raise many interesting questions: Why do their nuclei undergo decay? What are the products of such decay? What happens to the nuclei after they decay? And what can we learn about the structure of the nucleus from radioactive decay? These questions occupied Hahn, Meitner, and other researchers in the areas of nuclear chemistry and physics. Even today, a century later, we still do not fully understand these issues nor do we have precise answers to these important questions. We know that uranium decays because its nucleus is, in a sense, "too large"—it is packed with "too many" neutrons. But smaller nuclei undergo radioactive decay as well. The reason for this decay is not sheer size but rather the way the nucleus is packed.

Empirically, scientists now know about the existence of mysterious magic numbers: 2, 8, 20, 28, 50, 82, and 126. Elements with protons and neutrons totaling these numbers tend to be more stable than others and are less likely to decay. The reason for this phenomenon remains a mystery. It is not simply the "classical" packing of spheres—all micro-phenomena are given to the laws of quantum mechanics. The reason is deep in some quantum mechanism at present not well understood.

THE MYSTERY OF uranium decay and other nuclear processes was intrinsic to the Hahn-Meitner research team's goal of discovering actinium's origins. From what they knew about this process at the time, Hahn and Meitner believed that uranium did not decay directly into actinium, that there had to be some other, still-unknown, intervening element which later gives rise to actinium. (In fact, uranium decays into various elements, depending on conditions.) The "mother element" of actinium, they realized, could be a very weak radioactive element, which would necessitate very sensitive detection methods. To minimize interference in the radiation measurement by outside sources—such as traces of radioactive elements left over from previous experimentation—would require much care in maintaining special conditions in the laboratory.

It was of paramount importance that background radiation in their laboratory be kept at the lowest level possible. This led them to be obsessive about not importing impurities into the lab. Their lab assistants were required to wash often, keep their lab clothes and coats as clean as possible, and—something that was contrary to usual German custom—never shake hands, for fear of contamination. Rolls of toilet paper were kept next to public phones and all door knobs so that they could be wiped clean every time they were touched by anyone.[3] These extreme measures, designed primarily to increase the precision of laboratory work by reducing any stray radiation, had very fortunate consequences for Hahn and Meitner. For unlike the Curies, Enrico Fermi, and other scientists involved in experiments with radioactive substances, Hahn and Meitner thus kept the radiation exposure to their own bodies to low levels and, as a consequence, each lived a long life. (They were both 89 when they died, an unusually advanced age for scientists handling radioactive materials.)

Actinium was extremely scarce; it was found only in pitchblende or other uranium deposits, and, as noted, it had a relatively short half-life (Hahn and Meitner estimated it at 25 years; today we know it is closer to 22). Because of this short half-life, actinium had to be constantly replenished or else it would disappear forever, so it had to have a "mother" that was present in the pitchblende and was itself derived from uranium via the decay of that metal. But Hahn and Meitner concluded that the uranium decay leading to actinium had to be a different sequence of nuclear processes from the uranium process that results in radium and radon. Actinium was not hard to detect chemically, and when uranium was studied, it never was seen to lead to actinium—which is what led Hahn and Meitner to the understanding that there was something mysterious going on.

At the time, there was much confusion and incomplete information about radioactivity and its processes, and about the attendant problem of how to place newly discovered elements within the periodic table. Scientists also were not sure whether Mendeleyev's table could actually maintain its structure at the periphery of human knowledge about the elements. Could the simple linear progression discovered by the Russian chemist in the middle of the previous century hold true for elements that were radioactive, and hence intrinsically unstable, decaying to other elements over periods of time lasting anywhere from seconds to many thousands of years?

Since some elements decayed extremely slowly, they were not always identified as radioactive at all, while others, decaying too fast, could not be identified chemically before they had changed into something else. The science of radioactivity was new, and discoveries were made every day. But to put all the new knowledge and hypotheses and guesses together within a consistent theory was a major challenge. In 1913, however, the Polish nuclear chemist Kasimir Fajans and, independently, the Scottish scientist Frederick Soddy discovered an important property of radioactive decay. They realized that when a radioactive element undergoes alpha decay—that is, once it emits a helium nucleus—the product is a chemical element that lies two places below it in the periodic table. For example, when thorium (atomic number 90) undergoes alpha decay, it becomes radium (atomic number 88). Also, the two scientists working independently had discovered that when a radioactive element emits a beta particle (an electron), it becomes an element that is found one place above it in the periodic table. For example, actinium (atomic number 89) undergoing beta decay will give rise to thorium (90).[4]

These newly discovered laws explaining the relationship between chemical properties and the nature of the radioactivity emitted showed that the mass was independent of chemical activity—the latter determined solely by the atomic number, and not the atomic weight. Frederick Soddy coined the term isotopes (from the Greek for "equal place") to denote two or more elements that had the same place in the periodic table but had different radioactive properties and different weights. In other words, isotopes of the same element are chemically identical but have different atomic weights.

These discoveries led to the realization that there were various groups of elements within the periodic table. Hahn and Meitner hypothesized that actinium belonged to what was denoted as Group III. If this was indeed true, Hahn and Meitner deduced, then because of the laws of alpha and beta radiation that had just been discovered, there had to be a gap in the table in Group V, in which there was an as yet unknown element whose place was between thorium (atomic number 90) and uranium (atomic number 92). That unidentified element (atomic number 91) had to be the "mother element" of actinium. But where was it? How could they find it?

The mysterious new element might be very weakly radioactive but it had to exist somewhere inside pitchblende or some uranium salt, because extremely minute quantities of actinium were always found in the presence of uranium. Hahn and Meitner obtained two sources of uranium. One was an old sample of a uranium salt, uranium nitrate, that had been used in their lab for some years, and the other was a sample of raw uranium ore—pitchblende from Joachimsthal. Both sources led to a mélange of decay products of uranium: elements in the tantalum group as well as polonium, thorium, radium, plus other elements that are produced as uranium undergoes various stages of radioactive decay. All these products had to be observed and studied chemically and electroscopically. The logic behind their lab procedure was that the radiation from the short-lived isotopes created from uranium decay would fade over years of observation, while the actinium (whose half-life they estimated at 25 years) would build up faster. Eventually, the presence of the (perhaps weakly radioactive) element leading to actinium would become evident, and then chemical analysis could lead them to finally identify it. This challenging detective work had to be conducted very carefully over several years.

IN THE SUMMER of 1914 Hahn and Meitner were right in the middle of the most promising part of their research. They were developing new chemical procedures using silicon dioxide to remove tantalum products from the mixture of isotopes in pitchblende, procedures that made it easier to search for the mysterious "mother" of actinium. A calamitous global event interrupted their search as World War I exploded. Otto Hahn, who was a member of the German army's reserve units, was called to action and had to leave the Berlin laboratory.

Hahn was eager to go to war and fight for his country. The German-speaking nations—Germany and Austria—believed that they were in the right, and even Meitner, while hardly excited about the war that raged through Europe, seemed to support her native Austria's position. It should be noted, however, that the acceptance of the war was not absolute among scientists: Einstein, in one glaring exception, opposed the war in the strongest terms, and his hatred of all wars and the military in general are well known.

While Hahn was on the front in Belgium, Meitner volunteered to help tend the wounded in field hospitals in the east, where she often saw dying Hungarian soldiers for whom little could be done medically. As the war progressed, the Germans embarked on a secret project to produce chemicals to kill the enemy at the front. Parts of the Kaiser Wilhelm Institute of Chemistry were covertly converted into labs for research on the production of poison gas. Meitner—a female Jewish Austrian and hence not trusted by the authorities—was kept out of these labs and left in the dark about this effort. From her letters, however, it is evident that she had "a fairly good idea" about what was going on in the Institute.[5]

Otto Hahn was designated a chemical warfare specialist and ordered to participate in experiments using poison gas at the front in Belgium. These were the first instances of the use of chlorine gas in warfare and the results were horrifying. Enemy soldiers exposed to chlorine in the trenches died in dreadful agony, a fact well-known to the German scientists conducting the experiments, whose group became known as the Pioneer Regiment. It had been organized by the chemist Fritz Haber, and in addition to Hahn it included several other German scientists, James Franck and Gustav Hertz among them.

In April 1915, this unit was transferred to the east, where it directed its gases against Russian soldiers on the front. The German scientists had changed the chemical to make it more potent and deadly—instead of chlorine gas, they used a mixture of chlorine and phosgene. Hahn and his co-workers could see the horrible effects of their creation on the Russian soldiers, who were dying in the worst pain ever seen in warfare. But he stayed with the unit until the end of the war, experimenting with various chemicals to make the poison gas as effective as possible. He excused his behavior in his memoirs by saying that work with noxious chemicals had "numbed" the minds of the German scientists so that they "no longer had any scruples."[6] This ridiculous excuse places in question Hahn's behavior during the Second World War, when his work on uranium would become important in warfare. One wonders whether he was then a dedicated Nazi scientist trying to help Hitler obtain an atom bomb or an innocent scientist whose work to advance scientific knowledge was used against his will in something far more sinister.

In January 1917, while Hahn was on leave in Berlin, he and Meitner continued work on the experiment that had been dormant for long periods during the war. For four years, the tantalum-group sample they had extracted from the uranium nitrate had not yet shown any sign of yielding actinium, so the pair felt that the probability that the mother element of actinium would be discovered in it was small. They discontinued that experiment and turned their attention instead to the silica they had extracted from their pitchblende sample. The silica contained all the tantalum-like elements from the pitchblende. When they analyzed what was now in this mix of compounds that had been radiating away for four years, they discovered what they thought might be a tiny amount of actinium!

While they were working feverishly trying to isolate a microscopic amount of actinium from the mixture—an effort that took many months—organizational changes were announced by the administration of the Kaiser Wilhelm Institute of Chemistry. The institute divided the Hahn-Meitner Laboratorium into two separate departments: one part was a physics section to be headed by Lise Meitner; the other part remained a chemistry lab to be directed by Hahn. With the new appointment, Lise Meitner received a raise, so that her salary was on par with that of Hahn.[7] While now they each had a separate department to head, Hahn and Meitner would continue to work together whenever he was in Berlin.

When Hahn was at the front working on poison gas, Meitner continued their experiments. She pulverized another 21 grams of pitchblende and boiled it in nitric acid; this was to give her more residue of silica. This process yielded 3.5 grams of residue with the same tantalum-like elements—and hence the putative actinium "mother" among them—with which to work. She set aside 1.5 grams as a control element, and worked with the remaining two grams. Further steps followed, including a treatment of the residue with hydrofluoric acid, the most potent solvent. From that solution, various elements that release alpha radiation were very slowly separated, until Meitner was left with one element only—a slow alpha emitter, the suspected "mother of actinium." This remaining element now had to be monitored for a reasonably long period of time to see whether its radioactive decay actually yielded actinium.[8]

Several delicate problems remained. The scientists had to be sure that actinium was observed without its own decay products—that is, the element to which actinium was decaying had to be removed by some means. Also, one had to be able to determine the difference in the alpha radiation from actinium and from its mother element. In April 1917 Hahn was again at the lab for a short leave, and the excitement of feeling they were now so close to reaching their goal after so many years consumed them both. They worked round the clock. They knew that the chemical element they were after was sitting right in front of them in the silica residue distilled from the pitchblende, but they had to isolate it completely.

The preparation was carefully monitored for another six months, and Meitner observed and measured exactly what she had been expecting: a linear growth in the radiation that was characteristic of actinium, hand-in-hand with a decline in the radiation that seemed to characterize the precursor of actinium. She then proceeded to measure and estimate the half-life of the mother element. In the fall of 1917, when Hahn was in Berlin, the two continued to isolate the new element and estimate its parameters. In March of the following year, the pair sent out a paper that was the result of years of research under the stress of war and famine, conditions that did not favor scientific work. This was their triumph. The paper, entitled "The Mother Substance of Actinium, a New Radioactive Element of Long Half-Life," was published in the premier German physics journal. In it, Hahn and Meitner suggested the name *protactinium* (from "proto-actinium," parent substance of actinium) for the element they had discovered. While Meitner had done much of the work alone, Hahn's name appeared on their paper as senior author.[9]

Thus in 1917, after a long series of arduous experiments, Meitner and Hahn had finally made a major discovery: They had isolated the radioactive element protactinium—a metal chemically similar to vanadium and found in tiny amounts in uranium ore. They had applied the work of Soddy and Fajans to find a new element, and in doing so, added weight to these researchers' conclusions. This was an important achievement, but the pair knew that much work was still ahead. They had yet to answer key questions about precisely how uranium turned into other elements: What was the actual process of radioactive disintegration? This remained the deep mystery that underpinned the discovery of their new element. Unbeknownst to them, they had fierce competitors searching for the same answers in both Paris and Rome.

IN THE DAYS before the Internet, scientists collaborated in various ways. Many conferences were held at different levels, some with participants from around the world, others more limited by narrow discipline or geographical location. Travel across the oceans entailed long sea voyages and thus participation at a meeting in another continent was difficult. But in Europe, scientists could meet with relative ease because of an efficient rail system (which is still excellent today). Papers sent to journals were also distributed as preprints, so scientists could learn quickly what others in their field were doing. And finally, scientists would spend a year or two at a leading laboratory or center in countries other than their own to learn techniques from the great masters, such as Ernest Rutherford and Niels Bohr.

IN FRANCE, IRÈNE CURIE, Marie and Pierre's eldest daughter, interrupted her studies at the Faculty of Science in Paris to serve as a nurse and radiographer during the First World War, teaching field doctors at the front how to operate the new X-ray machines. She earned the nation's accolades for her heroic contribution to the war effort. After the war, she returned to complete her studies. She shared her mother's features (a steep forehead, high cheekbones, and naturally curly, light brown hair), and she spent so many hours alongside Marie (for whom she had deep admiration) at the Radium Institute that lab assistants referred to her as "The Crown Princess." Irène began her own work on radiation to qualify for a doctorate in physics from the Sorbonne. When awarded that degree in 1925, the event was reported in the world press. She was destined for a career in science, and the Parisian press often reported on the achievements of this rising star who shunned convention, dressed in baggy clothes, and avoided social pretense.

As the war had ended, Marie Curie had interviewed possible new staff and hired young Frédéric Joliot, whom Paul Langevin had recommended. Before long, Frédéric was working with Irène on uranium, and after a few years of fighting with each other over who was the leader of their joint experiments, the two married. One of the things they had in common was an almost unnatural admiration for Marie. As an adolescent, Frédéric had clipped newspaper reports about the elder Curies' discoveries and posted them on his wall. Now Frédéric and Irène were determined to carry on Marie's earlier work and expand it.

In pursuing this larger goal, the new research team of Curie and Joliot was intent on exploring exactly the mystery that Meitner and Hahn were determined to solve in Berlin. The specific question— which they hoped could be generalized to problems about radioactivity in general—that occupied both teams well into the 1920s and 1930s was: What happens to uranium when it is bombarded by particles? Neutron radiation (the release of neutrons from radioactive nuclei), as we've seen, was discovered in the early 1930s in England by James Chadwick (1891–1974), and physicists wanted to know what happens when a uranium nucleus is hit by a neutron.

In particular, if uranium—the heaviest natural element—could absorb that additional neutron, would it form a larger, heavier "transuranic" atom, which some scientists hypothesized existed? (An example of a transuranic element is plutonium, whose atomic weight is 239 and is thus heavier than uranium 238, but plutonium is an artificial element created through a different process and would not be discovered until much later.)

In 1932 in Rome, a solo genius began to work on this same problem. Within a few years, he would become convinced that he had an answer—experimental proof that could preempt his competitors in both Berlin and Paris.

LABORATORY WORK IN physics and chemistry was a very arduous occupation in those days—as it still is today. It required long hours of preparation of the reactive compounds to be used in an experiment. For example, raw ore might need to be treated by an acid or an oxidizing agent (something that acts like bleach) for hours, days, or weeks—and would have to be closely observed at given intervals during the period of preparation to ensure that it was not overtreated. Then there were the experiments themselves, which had various stages, each of which could be very lengthy in duration. Finally, calculations had to be made, mathematical modeling using equations had to be derived, and the results had to be analyzed and recorded. This was exhausting work that, literally, never ended. The quest for knowledge in the lab often meant that everyone sacrificed time and part of their personal lives to the demands of the research.

In the period between the two world wars, governments invested heavily in research and development. Industry had to be rebuilt and the economy put back on its feet. The Germans especially, but also the British, French, and Americans, invested in educational and research institutions. Science was making great progress, and as today, some areas were more popular than others and drew the sharpest minds. This is what happened with physics generally and with radioactivity research in particular.

Scientists devised clever experiments, which they then speculated about and analyzed using a combination of disciplines, such as chemistry, mathematics, and theoretical physics. These state-of-the-art scientific problems prompted different teams to engage in a fierce race to uncover truth.

5

ENRICO FERMI

Enrico Fermi (1901–1954) was born to a middle-class family in Rome. His father, Alberto, a self-taught man who came from an agrarian background in the Po Valley, was a high-ranking official of the Italian railroad system. His mother, Ida, was a schoolteacher from Bari, in the south of Italy. The family lived in an apartment with only cold water near the railroad station. Their three children were close in age: The oldest was Maria, born in 1899; there followed Giulio, born in 1900; and the youngest, Enrico, born in 1901.

Enrico and Giulio were extremely close—so close, in fact, that they had no other friends. The two boys shared everything, keeping secrets, building ingenious machines, and solving math and physics puzzles. But at fifteen Giulio died tragically from mishandled anesthesia while undergoing a simple outpatient operation to remove an abscess from his throat. This tragic loss was a terrible blow to the family. Giulio was Ida's favorite child, and after his death she was inconsolable. Still swept by that continuing sorrow, she died when Enrico was in his mid-twenties. Enrico lost not only a brother but also his close friend, and he withdrew from the world to find solace in books.

Like Lise Meitner, Enrico Fermi knew by age fourteen that he wanted to be a physicist. But unlike the Austrian, who had to overcome cultural blocks due to gender, the young Fermi had little trouble pursuing his passion. He loved two things above all else: hiking and reading. In Rome, he could find any book he wanted on physics and mathematics. The nineteenth century had been a heyday of Italian mathematics and theoretical physics, and in earlier centuries Italy had made advances in science in the grand tradition of Galileo. As a teenager, Enrico found books by the best Italian scientists and devoured them. Finding these books, moreover, was a pleasant adventure he refined to an art. Every Wednesday, market day in Rome, he would go to the Campo dei Fiori, the square in the center of the city where Giordano Bruno was burned at the stake in 1600. Many stalls selling used books were erected for the market, and Enrico would search through them for anything of interest. It was here in 1915 that he obtained the volumes of *Elementorum physicae mathematicae* (Elements of Mathematical Physics), written by the Jesuit Andrea Caraffa in 1840, seventy-five years earlier.[1] The fourteen-year-old already knew enough Latin, which was required in school, to read and understand this book. This was the young boy's first physics book; many others would follow from the same source. While reading the book, Enrico commented to his sister: "It's wonderful! It explains the motions of the planets!" and "I am learning about the propagation of all sorts of waves!" Maria tried to feign interest, but it was difficult. She wasn't interested in physics or mathematics. One day, Enrico told her: "Do you know, it's written in Latin—I hadn't noticed."[2]

Fermi was distinguishing himself in mathematics and physics at his school, and had a thirst for knowledge satisfied only through advanced study on his own. Around that time, he had a conversation with Adolfo Amidei, a railroad official who worked under Fermi's father. He knew that Amidei, who was then in his late thirties, was an amateur mathematician who passionately pursued reading and solving problems in mathematics and physics. He also knew that Amidei had an extensive library. Enrico would often come to meet his father at the railroad offices and accompany him home. One day, he began a conversation with Amidei about mathematics. Fermi asked Amidei: "Is it true that there is a branch of geometry in which important geometric properties are found without making use of the notion of measure?"[3]

Amidei responded that there was indeed such an area of mathematics, and that it was called projective geometry. He offered the teenager a book on the subject, which Fermi finished in a short period of time. To Amidei's surprise, he found that not only did Fermi understand everything in the book, but he even solved all of the several hundred problems in the text—some of which Amidei had struggled with—with no difficulty. Then Fermi asked Amidei if he could borrow more books. The stunned Amidei recognized that his boss's son was a remarkable genius. As he would later write to Fermi's future student and fellow Nobel Prize winner Emilio Segrè, "I became convinced that Enrico was a prodigy, at least with respect to geometry. I expressed this opinion to Enrico's father, and his reply was: Yes, at school his son was a good student, but none of his professors had realized that the boy was a prodigy."[4]

During the war years of 1914 to 1918, Amidei nurtured Fermi's intellectual development, generously lending him many books on mathematics and physics from his own library, encouraging Fermi to keep each book for a year or longer, and discussing the books' ideas and problems with the younger man. The best mathematicians and physicists of the day and the previous century had written these textbooks. The book on algebra was written by Ernesto Cesàro, the one on analytic geometry by Luigi Bianchi, the calculus text by Ulisse Dini, the book on logic by Giuseppe Peano—these authors comprise a veritable "Who's Who" of Italian mathematics at its pinnacle. There were also books by other prominent European mathematicians such as a German text by H. Grassmann, and a major treatise on theoretical mechanics (which, interestingly, was taught in those days as part of the mathematics, rather than the physics, curriculum) by the eminent French mathematician Simon-Denis Poisson. Fermi could read these texts with little difficulty: French was taught at school, and he taught himself enough German to understand mathematical texts and papers. The budding mathematical genius assimilated huge amounts of information far beyond what was taught at the gymnasium which prepared students for the university.

While Rome had, and still has, an excellent university, Amidei suggested that Enrico be sent to the Scuola Normale Superiore in Pisa.

The Scuola Normale Superiore was the most prestigious college within the University of Pisa, which had, over the centuries, attracted the best mathematicians, starting with Galileo in the seventeenth century, followed by two of the mathematicians whose books the young Fermi had by then mastered: Bianchi and Dini. The *crème de la crème* of Italian mathematicians worked there: Gregorio Ricci-Curbastro, Vito Volterra, and Guido Fubini, names known to all students of mathematics.

Having lost one child, Fermi's parents were reluctant to let Enrico move away, especially since Rome had its own excellent university. But Amidei pleaded with them to let their son go to the best mathematical and physical institute he knew. He stressed that the boy's future depended on the quality of his education, and after a few months was able to convince them to relent.

While there were ten thousand students at Pisa, the Scuola Normale Superiore enrolled only forty. The select group of students was housed in a Renaissance palace, the Palazzo dei Cavalieri ("Palace of the Knights"), close to the Leaning Tower of Pisa.[5]

Despite the high level of the studies, academic accomplishment came easily to Fermi. In 1919, during his second year at the university, Fermi began to keep a notebook in which he recorded everything he learned about physics—in class and through his independent readings. He developed entire theories in this notebook, which remained with him long after he had finished his studies and left Italy for the United States. (It was one of the documents among his papers that were bequeathed to the University of Chicago on his death in America in 1954.) In the notebook he recorded his understanding of the first elements of the quantum theory—blackbody radiation analyzed by Max Planck, Einstein's results on the photoelectric effect, and Bohr's concept of the atom—as well as the knowledge he had acquired of relativity. Both his professors and fellow students soon realized that he was a young man with truly exceptional abilities, a genius of the caliber of the great physicists whose works he was studying on his own.

Fermi soon emerged as the leader of the physics department in the school and the university. One of his professors teased him whenever he saw him, "Fermi, teach me something."[6] He was allowed to go to the physics lab whenever he chose—a privilege never given to any other student in the school's history—and to design his own experiments. It was during this period that a key characteristic of Fermi's intellectual philosophy became clear: He combined perfectly the intellectual qualities of both theorist and experimentalist. His rare mathematical abilities allowed him to develop physical theory exceptionally well, and at the same time, he pursued innovative experiments, always building his own lab apparatus, an accomplishment he valued. He continued in this vein throughout his career: He would write very abstract papers on physical-mathematical theory and build complicated experimental tools on his own in order to test and amplify the theoretical results.

But it was not all hard work and deep thinking. Fermi and the other students of his group spent much of their free time hiking in the Apennine Mountains, which run the length of Italy and come close to the Ligurian coastline near Pisa. While the Apennines are much lower than the Alps in the north of the country, they rise rather steeply from sea level in the region of Pisa. The famous quarries of Carrara, producing what is arguably the finest marble in the world, used by Michelangelo, Donatello, and other great sculptors throughout history, are located here. Above these quarries rise very steep mountain slopes that Fermi and his friends loved to climb on weekends. He became a strong climber, leading every hike up the mountains, constantly pushing himself higher, determining when and how long his friends should rest on the way up. He used to claim that his short legs were the strongest of all and could go for hours without the need to rest.

Like many young people, the students at the school liked to play pranks on unsuspecting citizens. They founded something they called the "Anti-Neighbor Society," and would devise what they thought were clever practical jokes. In those days, Italy had public urinals on some street corners. The students would steal small pieces of sodium from the lab and toss them into the urinals while men were using them. The sodium would explode when it hit the water, scaring the men half to death.[7] This irreverent attitude extended to the lab itself.

In chemistry, the students were handed samples of compounds to be analyzed using lengthy procedures that subjected the compounds to various reactive agents. The process and results were to be noted and would ultimately lead to a final identification of the compound in question. Fermi, who hated wasting time, discovered that looking at the compound under a microscope was the easiest and fastest way to identify it—the compounds were often cheap chemicals bought at cleaning supply stores, and he knew what they looked like under the microscope. He let his friends in on this deception at the lab. After a quick identification, he and his fellow students described in their reports the complex identification procedures they never undertook.

Fermi had a close friend in the department, Franco Rasetti, with whom he hiked and perpetrated pranks. Rasetti convinced Fermi to spend hours lying in fields to hunt geckos and lizards. They would then release the animals at the student cafeteria to scare the country girls who worked there as waitresses. This became a frequent diversion for the two young men. According to Enrico's wife, Laura Fermi, Enrico spent the "hunting time" thinking about physics problems while on the grass, his eyes glued to the ground in search of these reptiles. She wrote, of one such session several years later:

> While he watched the ground, ready to pull his lasso should a gecko appear, he let his mind wander. His subconscious worked on Pauli's [exclusion] principle and on the theory of a perfect gas. From subconscious depths came the missing factor Fermi had long sought: no two atoms of a gas can move with the same velocity or, as physicists say, there can be only one atom in each of the quantum states possible for the atoms of a perfect monatomic gas.[8]

In 1925 the Austrian physicist Wolfgang Pauli formulated his famous "exclusion principle," one of the quantum rules governing the behavior of particles in the micro world of atoms. The Pauli exclusion principle says that no two electrons (or other *fermions*—certain kinds of particles later named after Fermi) in an atom can have all their quantum numbers equal. They have to differ in at least one aspect; for example, if they are in the same orbit, they must have opposite spins.

Fermi applied Pauli's principle to a large ensemble of elements: a gas made of single atoms. He was thus able to show that, under special circumstances, a quantum principle could also be applied to a larger system. This was an important discovery made while indulging in juvenile folly, because it provided an example in which a quantum rule extends to entities that are much larger than a single electron or proton or atom—to collections of particles making up a totality that can actually be seen by the human eye: the gas inside a bottle, for example. Major work in this area had already been done by the Indian physicist S. N. Bose and by Einstein in 1924, leading to what is known as a Bose-Einstein condensate, a collection of particles that exhibits, as a whole, quantum behavior. Fermi's work was an important step forward in this same direction.

Fermi distinguished himself as a superb theoretician and experimentalist. One of the key experiments he designed at the physics lab was an X-ray analysis of crystals, which revealed the Laue diffraction patterns of the crystal lattice. Here, too, Fermi had actually built his own X-ray tubes for the innovative experiments that led him to important new results. This was his first foray into the realm of radiation, and he found he was attracted to nuclear physics.

In a seamless continuation of his undergraduate work, Fermi wrote his thesis on radioactive processes and received his doctorate in physics in 1922 from the university in Pisa. By then he had published a number of papers, some theoretical and some experimental. That year he made an important discovery in general relativity, showing that space is Euclidean. The related paper that he published—which encompassed a mathematical theorem he derived to demonstrate the highly theoretical result that in very small neighborhoods of the universe, space is flat—established him as a leading scientist very early in his career.

The shape of space is one of the mysteries of science, and each new generation of physicists, astronomers, and cosmologists makes new discoveries in this area. Einstein's general theory of relativity showed that space is curved around massive objects. The theory actually defined mass in terms of the curvature of space. The inflationary universe theory, developed by the MIT physicist Alan Guth in the 1980s, shows that space is nearly perfectly flat on a very large scale (beyond local distortions by mass), and this has been confirmed by experimental work in cosmology (which correlated speed and distance of faraway galaxies) in 1998, and by satellite measurements of microwave background radiation in space.

Fermi was a man of method, as his wife later described him. His brain worked like a clock. He would set to work early in the morning, and continue without stop until 9:30 at night, when he promptly went to bed. He found all physics problems to be "clear" or "obvious." When faced with a problem, he would close his eyes and think for a few moments, then blurt out the answer as if deriving it had taken little mental effort.

Over the next few years, Fermi traveled to several universities in Europe. In 1923 he spent some months on an Italian government fellowship in Göttingen, a prestigious university in Germany, working with quantum pioneer Max Born. In 1924 Fermi spent three months on a Rockefeller fellowship at the University of Leiden in Holland, working with physicist Paul Ehrenfest, Einstein's good friend. After his return from Leiden, Fermi took a position in the physics department of the University of Florence. The physics buildings were separate from the rest of the university, outside the city of Florence in the village of Arcetri, where Galileo was confined in his villa by the Inquisition from 1633 until his death in 1642.[9]

Fermi's friend Rasetti was also in Florence and the two worked together enthusiastically. Fermi taught Rasetti much about theoretical physics while the latter worked out ingenious experimental designs for their joint work on mercury resonance in weak magnetic fields, obtaining good results and making important contributions to atomic spectroscopy. This collaboration encouraged Fermi, and he became more and more concerned with turning his fast-accumulating research papers on important topics at the frontiers of modern physics into something that would give him a full-time academic position that would allow him to teach, which he loved, as well as to continue advanced research in physics.

In 1925, Fermi applied for a position in the physics department at the University of Cagliari in Sardinia, but he was turned down because a majority of professors there did not believe in Einstein's relativity theories, about which Fermi had written widely. Two famous

professors there, the mathematicians Vito Volterra and Tullio Levi-
Civita—whose results in absolute differential calculus were used by
Einstein in devising his general theory of relativity in 1915—voted for
Fermi, but the other three prevailed and another candidate was ap-
pointed.[10] Fermi remained in Florence.

He was much more successful in applying for a position in Rome,
at a significantly superior institution. The chair of the University of
Rome physics department was Orso Mario Corbino, who was also a
senator (a lifelong member of Italy's upper house in the monarchy)
and thus involved in the political life of the country. He was intent on
creating the best physics department in Europe. In the fall of 1926, a
national competition was announced for the university's chair of theoretical physics. The selection committee met on November 7, 1926,
and made its decision: Enrico Fermi was chosen overwhelmingly. The
scientific work already completed by this twenty-five-year-old physicist was deemed of such superb quality that he was considered the
only candidate to deserve this position. Fermi was elated to be able to
return home to Rome and to indulge his dual loves of teaching and
engaging in leading-edge research in physics.

Corbino had the vision and insight to recruit the greatest young
physicists of his day, and Fermi was to be his rising star. In 1927
Rasetti was invited by Corbino to join Fermi's team in Rome and he
happily came from Florence to work with his friend. While hiking
with Rasetti in the mountains, Emilio Segrè, then an engineering student in Rome, learned about physics through long conversations with
the young professor and quickly decided to change his major to
physics and to study under Rasetti's friend, the famous Fermi. He had
once heard Fermi give a public lecture on the new quantum theory
and became enthusiastic about it. Segrè became Fermi's first and
brightest doctoral student and was incorporated into the group
Corbino was assembling.

Some months later, Segrè convinced his friend Ettore Majorana,
then also an engineering student in Rome, to switch majors and to
join the emerging physics team. Majorana was a genius of unparalleled
abilities, a brilliant mathematical physicist. He was the only person
who could surpass Fermi in speed of mathematical computations,

some of which he would do in his head while others required paper and pencil or some of the early calculating devices (slide rules) available in the 1920s. Majorana's life would end early in a mysterious disappearance, perhaps a suicide at sea, in 1938.

The physics department was housed in a magnificent old palace on the Via Panisperna in Rome, somewhat separated from the other buildings of the university. Corbino's group of brilliant young physicists quickly acquired a degree of fame within the university and the community and became known as "the Boys of the Via Panisperna."

Fermi and Rasetti remained very close and developed mutual habits. They would both speak in very low voices with deliberate, slow enunciations. This was a style that was emulated by the other young physicists in this growing group of mavericks. Emilio Segrè tells a story about a physicist traveling on a train in Italy at that time and making conversation with a fellow passenger. His companion turned to him and asked: "Are you by chance a physicist at the University of Rome?" "Yes!" replied the member of Corbino's group, "how did you know?" "Ah, from your way of speaking," was the answer.[11]

ON JULY 19, 1928, a sultry day in which the temperature in Rome reached 104, Fermi married Laura Capon. The two had met in the spring of 1924, when she was 16 and he 23, through common friends with whom they played soccer. They later hiked together in the Dolomite Mountains in the summer of 1926. When Fermi became a full professor of theoretical physics in Rome the following winter, Laura was a general science student at the university. She wasn't taking any of the classes taught by the famous young professor, but the two met again at the house of a mutual friend, a mathematician at the university. After a courtship of a couple of years, they decided to marry. Because Fermi was Catholic and Capon was Jewish they could not have a religious wedding, so a civil ceremony had to suffice, and they married during what turned out to be the hottest day of 1928, at the Rome City Hall, located on the Campidoglio, the hill that was the heart of ancient Rome.

After their wedding, the Fermis moved to a small apartment in the center of Rome, not far from the university. It provided them with just enough creature comforts. They had a small convertible Peugeot, which they drove in the countryside around Rome as well as to Florence to visit Laura's uncle, and continued to enjoy hikes in the country and vacations in the mountains. For their honeymoon in 1928, they became one of the first Italian couples to fly commercially. They took a flight on a hydroplane from Rome to Genoa and continued by train to the Swiss Alps, where they vacationed for a few weeks. But their life was not without more serious concerns. Fermi, who was ambitiously eager to earn more money now that he had a wife to support, recruited his young bride to help him write a textbook on atomic physics. He wrote it all from memory—without the use of any books or notes—while in the Alps and later in their small apartment; she edited and typed it for him. The textbook provided them with additional income for many years.[12]

In 1929 Enrico Fermi was elected to the Royal Academy of Italy, which Benito Mussolini had just established with his customary pomp. Its members wore feathers and swords and had elaborate uniforms—all of which Fermi abhorred and refused to don whenever possible. Each member had a title and was addressed as "*Eccellenza*" ("His Excellency"). The Royal Academy was envisioned by Mussolini in 1926 as the academy to end all academies—it was intended to showcase fascist science in a world that had not yet seen the true, ugly face of fascism. One physicist was to be named to the academy, whose members came from all the sciences. Many people had high hopes of a nomination—physicists much older and more established in their careers than anyone in Corbino's group. As a senator, by law Corbino himself could not be nominated. In March 1929 the final, secret selection of the new members of the academy was made public. Fermi was chosen to represent physics.

Fermi had hardly expected the nomination. But despite his antifascist feelings, he was elated. As his wife described it, the reason was purely monetary. Fermi wasn't paid well as an academic, and he always felt the need to supplement his salary with book royalties and lectures. But Mussolini paid the members of his new academy an additional income worth one and a half times Fermi's salary. This was good money and it pleased Fermi. He became known as "His Excellency Fermi," and his wife tells the story that they once traveled to a ski hotel in the Alps, and when they checked in, the hotel owner asked him, "Are you somehow related to His Excellency Fermi?" "He's a distant cousin," Enrico replied, and the couple was left to ski in peace without being gawked at by admiring strangers.[13]

Fermi was known for his sense of humor. One day, he was to address a special meeting of the Italian Academy of Sciences held in Mussolini's Palazzo Venezia (from its balcony overlooking the wide Piazza Venezia, Il Duce would often give his fiery speeches). Of course, the palace was ringed by the Italian state police, the *carabinieri*. All the other members of the academy arrived at the Palazzo driven in limousines and were quickly admitted. But Fermi, notoriously averse to being driven by anyone, came in his little car. Since he drove an unimpressive automobile, he was immediately stopped by the *carabinieri* guarding the palace and asked who he was. He was worried that if he told them "I'm His Excellency Enrico Fermi," they would never believe him. So he said, "I am the driver of His Excellency, Signore Enrico Fermi," to which they replied, "Fine, drive in, park, and await your master."[14]

IN 1930 FERMI became the first and the only Italian physicist to be invited to the Solvay Conference, the sixth such meeting since these symposia had been inaugurated in 1911. Here he met the key physicists of the day, many of whom were working on nuclear physics and radiation theory, as well as quantum mechanics. In 1929 Fermi had published an important paper, "Quantum Theory of Radiation," that laid out theoretical underpinnings of radioactive processes. While working on this groundbreaking paper, which used results in quantum mechanics to explain what happens inside the nucleus as it disintegrates to produce radiation, Fermi had often consulted with the members of his group. He would perform theoretical calculations and then have them verified by Rasetti, Segrè, or Majorana. This was a very young group of scientists, ranging in ages from 20 to 28. Majorana, who would often verify Fermi's work in his head, was the only member of the group who could discuss highly theoretical mathematical concepts with Fermi on an equal level.

Spurred on by Corbino, the Fermi group realized in 1930 that theory was not enough and there was a great need in Rome for solid experimental work on radioactivity and associated processes. Since the Rome physicists knew less in this area than did many other physicists, they decided mutually to send members abroad to study with other groups doing advanced experimental work. Emilio Segrè was sent to study experimental techniques in Hamburg, and Franco Rasetti was sent to learn how to work on radioactivity with Lise Meitner in Berlin. The experimental techniques that Rasetti learned in Meitner's lab at the Kaiser Wilhelm Institute in Berlin and brought back to Rome would prove essential for the experiments on neutron radiation and uranium that were to be performed by the Rome group.

Enrico Fermi had spent the first period of his scientific career learning and making contributions to areas of physics that were growing during that time: relativity and quantum mechanics. He had trained himself to become a unique physicist, one acknowledged by his peers to have mastered perfectly skills in both theoretical and experimental physics. He was now ready to take on the most important scientific issue of the day: the study of the nature of radioactivity and its effects.

Wolfgang Pauli, along with Fermi and others, had been studying beta decay, the radioactivity emitted by certain elements, which consists of electrons emanating from the nucleus (rather than from the electron cloud, the orbitals around the nucleus where electrons naturally live). Pauli had been puzzled about the fact that a small amount of energy was always missing from calculations about beta decay. Since energy and momentum must be conserved in physical reactions, he concluded that something was lacking in their existing theories if the previously accepted law of conservation of energy was to remain valid. In other words, the total energy of the entire system before and after the emission of the beta electron must be the same, according to the law of conservation of energy. But there was less total energy in the system: the nucleus plus the emitted electron (that is, the system *after* the beta decay) had less energy than the nucleus before the electron left it (that is, the system *before* the decay). Yet, no other particle had been observed leaving the nucleus together with the electron. To account for the "missing" energy after beta decay, Pauli proposed the existence of an uncharged particle so small, possessing minimal or no mass and rarely interacting with matter, that it was elusive and simply could not be detected! This particle, he suggested, was produced in the process of beta decay and accounted for the missing energy. At a 1931 physics conference at the University of Rome, presided over by Fermi, Pauli urged physicists to search for this hypothetical particle in their experiments on radiation.

Fermi, in 1934, concurred with Pauli's hypothesis, and incorporated this uncharged particle into his theory of beta decay, naming it the neutrino. He further posited that both the neutrino and its "opposite," the anti-neutrino, play an important role in beta decay. According to Fermi's theory, in this process, when a neutron decays, it gives rise to a proton and emits an electron and an anti-neutrino. In the reverse process, when a proton absorbs energy it can give rise to a neutron, a positron (the anti-particle of the electron), and a neutrino.

In 1956 the American physicists Clyde Cowan and Frederick Reines would make history by detecting the physical presence of the neutrino after analyzing the radiation emanating from a nuclear reactor in North Carolina. But in 1934, Enrico Fermi refocused his attention on the problem of uranium irradiation by neutrons.

6

THE ROME
EXPERIMENTS

E nrico Fermi embarked on an ambitious plan of systematically studying the mystery of uranium and the secret of radiation. In the 1930s,
he launched the first of the key experiments in his small laboratory at
the University of Rome. Fermi directed a highly motivated team of young
scientists who devoted all their time and effort to this scientific quest in the
leading edge of science. This innovative group's achievements, building on
the research conducted in Berlin, Paris, and Copenhagen, constituted an
important step forward in our understanding of nuclear reactions, and ultimately toward the practical use of atomic energy.

At the 1933 Solvay conference in Brussels, Fermi met all the great scientists involved in atomic and quantum physics: Paul Dirac, Wolfgang
Pauli, Louis de Broglie, Marie Curie, her daughter Irène Joliot-Curie, Ernest
Rutherford, James Chadwick, and Werner Heisenberg. Pauli's theory about
the missing energy in beta decay and the suggestion that another particle
might be emitted was discussed at the session. When Fermi returned to
Rome, he performed a deep theoretical analysis of the problem. Pauli's idea
was based on an educated guess rather than an experimental proof or a theoretical breakthrough. Working out the theoretical details of the problem,

and using all the experimental data available, Fermi was able to derive the actual law of beta decay, as presented in the preceding chapter. In addition to the two forces already accepted by physicists—gravity and electromagnetic force—Fermi's theory introduced a new force in nature, which he called the force of weak interaction; it plays a key role in some radioactive processes.[1]

Fermi knew that the theoretical contribution he had made to understanding the nucleus and certain kinds of radioactive decay was of great value, and he presented his results at a colloquium in Rome, impressing his listeners. He now planned to disseminate these ideas to the wider scientific world. Fermi thought, justifiably, that the journal *Nature* would be the best venue. But in one of its most embarrassing decisions in history, this prestigious journal rejected Fermi's paper. In his response, the editor wrote that he was rejecting the paper because it contained "speculations that were too remote from physical reality."[2]

Fermi had to be content with the publication of his results in an Italian journal and also in the German *Zeitschrift für Physik*. Fermi was working on his theoretical analysis of how beta rays suddenly emanate from inside the nucleus of certain elements and was preparing his paper for publication in the German journal when from Paris came a bombshell. Irène Curie and Frédérick Joliot had managed to achieve an equally great experimental advance in the field of radioactivity research.

IN JANUARY 1934 Irène Curie and Frédéric Joliot announced that they had been able to induce artificial radioactivity. Their results were published in the February 10 issue of *Nature* and caused a great stir. This was big news in science because until that time people thought radioactivity was something that occurred in nature and could not be induced by scientific experiments. Curie and Joliot were awarded a Nobel Prize the following year, 1935, for this discovery.

The Joliot-Curie team in Paris had exposed a thin sheet of aluminum to alpha radiation from a polonium source. Alpha rays are heavy particles—they consist of four little particles held together: two protons and two neutrons—and when these alpha particles hit other nuclei, *sometimes* they induce artificial radiation, as the Joliot-Curie group discovered. Polonium was the source of the alpha rays.[3] The target was the aluminum, which responded to the bombardment of its nuclei by alpha rays with an *induced* radiation. This radiation was in the form of positrons (positively charged particles with the same mass as electrons; the positron is the anti-particle of the electron). Joliot and Curie found that the resulting aluminum positron radiation decayed exponentially, as happens with naturally occurring radiation, but the half-life of this artificially produced radioactivity was very small: 3 minutes and 15 seconds.[4] They then found that boron, when exposed to alpha rays, responded similarly and decayed exponentially with a half-life of 14 minutes, and magnesium did the same with a half-life of about 2 and a half minutes.[5]

If aluminum, boron, and magnesium reacted that way, would other elements exhibit the same response when exposed to radiation? This was an important question because the answer perhaps held the key to the very nature of radiation. Throughout the world, scientists were drawn to this new and fascinating problem.

IN ROME, ENRICO FERMI worked on this problem, and he decided to make one of his many professional switches from theory to experimentation. He had been engaged in theoretical work for quite some time. Now he wanted to turn his attention to experiments so he could elaborate on and even surpass the Joliot-Curie discovery.

Fermi understood that alpha radiation was not very efficient in such experiments. The alpha particles are heavy and big; they can be stopped by a sheet of paper, which is something quite striking to see in the laboratory. Because they are positively electrically charged (the charge comes from the protons that make up half the mass of an alpha particle), they interact with the negatively charged electron cloud orbiting around the nucleus, which interferes with their trajectory. Then, those alpha particles that do make it to the vicinity of the nucleus can be repelled by the equally positively charged nucleus.

Therefore, only some alpha particles are actually absorbed by the target nuclei.

Fermi had a better idea. Rather than the naturally produced alpha rays, if he could somehow use the newly discovered neutrons (revealed just two years earlier, in 1932, by James Chadwick) as the attacking particles fired at the target nuclei, he could improve on the Joliot-Curie team's experiment design. The reason this would be better, Fermi concluded, was that neutrons are smaller and are electrically neutral. They could therefore slip by the electron cloud surrounding the nucleus without interference, and they would not be repelled by the electrically positive charge of the nucleus. In a sense, the unsuspecting nucleus could not defend itself against an attack by the small, fast, and electrically neutral "bullets" fired at it.[6]

Fermi knew from theoretical and experimental findings that if beryllium is exposed to a strong source of alpha particles such as radium or radon (a radioactive gas that we now know causes health problems when it seeps into houses from the ground), it responds by emitting neutrons. On its surface, this seemed like a wasteful project: It takes 100,000 alpha particles aimed at a beryllium sample to produce a single neutron emission. But Fermi was very advanced in his scientific thinking. He knew that even though a huge amount of alpha radiation was wasted in creating a single neutron, that single particle being neutral and unaffected by electrical charges within the atom would be highly likely to reach its target and hit a nucleus. It would be beneficial to use neutrons to bombard nuclei because the desired end result would be very efficiently achieved.

Fermi understood, in other words, that once he created freely flying neutrons through the process of radiating beryllium with alpha particles, he'd have a magic bullet in his hands. Once he had a neutron flying around inside matter, undeterred by the electron cloud and unrepelled by the nucleus, it would eventually smash right into the nucleus.

Fermi's thinking was very clever, and the world's scientific community followed his work attentively. Physicists at Niels Bohr's institute in Copenhagen tried to follow in Fermi's footsteps, as did others in laboratories elsewhere. Lise Meitner's nephew, Otto Frisch, a young

Austrian physicist then in Copenhagen, was in the habit of reading
Fermi's research papers and translating them from Italian to share with
his colleagues at the institute, all of whom followed the Italian maver-
ick's leads in research. Frisch recalled his amazement on learning of
Fermi's decision to work with neutrons: " . . . Enrico Fermi was
preparing to bombard elements, not with alpha particles (as we all
did) but with neutrons. That puzzled me, since neutrons were very
rare: you had to bombard beryllium and waste a hundred thousand
alpha particles to produce one neutron. What was the sense of using
such expensive bullets?"[7] But soon, both he and many other scientists
understood the shrewd move by Fermi—what seemed a counterintuitive waste was actually a brilliant way of creating the perfect "bullet"
to hit a nucleus.

In the University of Rome physics building, there was a very concentrated sample of radium, and Fermi chose it as the ideal alpha
source. When he placed a quantity of powdered beryllium inside a
tube of radon gas derived from the radium, he produced a source of
neutrons for the experiment. It was now time to prepare to test which
elements, other than the aluminum, boron, and magnesium, already
studied by Joliot and Curie, would produce artificial radiation as a response to an attack by neutrons.

The Rome group was not organized well for such state-of-the-art
experiments. The radon gas was obtained from the decay of a single
gram of radium (worth tens of thousands of dollars in today's values),
discovered almost by chance in the basement of the physics building,
and there was no ready-made experimental equipment to be found anywhere in Rome. The main device for working with radiation was a
Geiger counter, which, as noted, had been invented in 1908 by the German physicist Hans Geiger, an associate of Ernest Rutherford. It had
been improved twenty years later by Geiger and his student Walther
Müller so that it could detect all known forms of ionizing radiation.
Fermi could not procure this device—so he decided to build his own.

For this, Fermi wanted Rasetti, who was better than anyone at
building lab instruments and other technical devices. But Rasetti was
nowhere near Rome. He was in the midst of a personal crisis that
made him want to travel the world in search of meaning. At this par-

ticular moment in 1934, Rasetti was in Morocco and unreachable.[8] Fermi did what he thought was the next-best thing—he made his own Geiger counter and other apparatus needed for an advanced radiation research laboratory.

Fermi sent Emilio Segrè, his former student turned enthusiastic research colleague, to stores all over Rome. Every week, Segrè would go into the city with a large empty bag and would return later in the day with the bag full of chemicals.

The group in Rome then systematically conducted the experiments on induced radioactivity first carried out in Paris by Joliot and Curie. The chemical compounds Segrè brought to the lab were refined to yield pure elements. These, in turn, were bombarded by neutrons emanating from a strong glass tube containing compressed radon gas in liquid form, into which a beryllium strip had been added to prompt the production of neutrons. In this way, many of the elements in the periodic table were tested to see if they could be made radioactive for some period of time. But because the neutrons (and alpha particles) emanating from the source would also trigger the Geiger counter to measure radioactivity, the target elements had to be measured in a room other than the one in which they were being irradiated. This presented a problem. Many of the neutron-irradiated elements responded by becoming radioactive for very short periods of time—some of them for less than a minute before the radiation stopped. So Fermi and his colleagues would irradiate an element, and then run as fast as they could to the end of the corridor to another room where the Geiger counter was placed and the radiation measurement was made. This procedure was quite comical—Fermi racing down the corridor with an irradiated element in his hand many times a day.[9]

After they had tested most of the elements in the periodic table, Fermi and his colleagues reached the final known element: uranium. Now the most important—and most controversial—work began.

FERMI SUSPECTED THAT radioactivity and uranium were more directly linked than previously shown. He wanted to bombard uranium with neutrons, which, he hypothesized, would create transuranic elements (or transuranium elements, as they are now called)—that is, elements with atomic weights that are higher than that of uranium. Since uranium had a huge abundance of neutrons stashed away inside its bulging nucleus, held together by forces that bind the particles inside it, it was the heaviest element. Now imagine that you fire a neutron (itself obtained from a radioactive process that induces an element to emit neutrons) at this overfull uranium nucleus. Fermi assumed that this would make the uranium nucleus even larger, that the nucleus would somehow absorb the neutrons fired at it.

Fermi's hypothesis was both right and wrong. Uranium with weight 238, or U–238, can indeed absorb an additional neutron in certain circumstances, creating the transitional element U–239, which quickly decays to the then-unknown radioactive element Pt–239 we now call plutonium. These elements are often termed "artificial" or "synthetic" since the radioactive decay does not occur naturally.

But the much rarer isotope of uranium, U–235, which makes up only 0.7% of all naturally occurring uranium and could be found (in that small percentage of 0.7) mixed in with the rest of the unrefined uranium in the lab experiment, does not absorb a neutron it encounters. Even though the scientists all expected that a transuranium element would be formed, nature works quite differently, as was soon to be demonstrated.

FERMI HAD PREPARED himself for this research by spending the previous year traveling around Europe, visiting laboratories in which atomic research was being conducted, including Göttingen in Germany and Leiden in Holland. At these laboratories he learned how to perform an effective bombardment of uranium atoms by neutron radiation. Through reading physics publications, he was also aware of the work being done by his competitors in Paris and Berlin. Fermi was racing against them—he knew that both the Meitner-Hahn and the Joliot-Curie teams had been at it for years. What secret knowledge had they gained? Fermi was frustrated by his inability to plumb this hidden knowledge.

Sometime later, Fermi hired a man who would give him an edge over the competition: Oscar D'Agostino. Before he joined the Fermi team, D'Agostino had worked in Marie Curie's institute in Paris and was well versed in the details of the Joliot-Curie research. The Fermi team thus gained new insights into how to conduct experiments on radioactivity that would advance Fermi's new work on uranium.

The Fermi team's innovative experiments convinced them that transuranium elements were being produced by the neutron bombardment. But what the Fermi team still needed was a way to analyze the results, a method of observing what actually happened when a uranium nucleus was hit by a neutron. They had to definitively identify any transuranium elements that might have been produced by the reaction. This proved to be a very difficult, frustrating task; often, experiments simply failed without revealing much information. Fermi was losing hope.

Then in mid-1934, the team chanced upon what looked like the right experiment. The sensing devices used in the lab seemed to indicate that something heavier than uranium was being produced by the nuclear reaction. This caused immense excitement. Fermi concluded that he had now observed the elusive transuranium elements produced by neutron absorption by uranium, and he was ready to publish his findings.

He was right that the neutrons were taken in by the uranium nuclei. But he and his co-workers were wrong in believing that they had actually observed transuranium elements formed as a result of that absorption. They had failed to identify correctly what was being produced, perhaps because they were not well enough versed in chemical analysis. The Italian lab had excellent physicists, but lacked high-level chemical expertise.

After performing complicated experiments that bombarded uranium with neutrons from the radon-beryllium source, Fermi became convinced that he had indeed produced the first transuranium element, "element number 93." But this discovery was later disputed by several scientists, and today the consensus among scientists seems to be that element number 93, which was named neptunium, was first made in a lab at the University of California at Berkeley six years after Fermi thought he had made it. Fermi's "element 93" is now believed to have actually been a product with atomic number 91.

By late summer, Lord Rutherford in London had read Fermi's report and communicated it to the Royal Society. Publication soon made these results available to the Joliot-Curie team in Paris and to the Meitner-Hahn team in Berlin. If Fermi had proved that neutron irradiation caused the production of yet-unidentified transuranium elements, then it seemed that Fermi had beaten out his competitors. Irène Curie and Lise Meitner and their colleagues were puzzled: Could Fermi possibly be right?

IN 1935, IRÈNE JOLIOT-CURIE conducted her own careful analysis of neutron irradiation, which included a chemical analysis. She published a paper that claimed that one result was that the element lanthanum (a silvery white metal with atomic number 57 and atomic weight 139) was released by the experiment. But lanthanum was a lighter element than uranium (whose atomic weight is 238), and thus could not have constituted the putative transuranium product of Fermi's experiment, which was supposed to be heavier than uranium.

Hahn and Meitner were incredulous: Had not Fermi proved just the opposite? How could a lighter element than uranium come out of the reaction? Meitner concluded that Irène Joliot-Curie was wrong. A few months later, Fermi organized a conference in Rome. Both Otto Hahn and Frédéric Joliot attended while their partners remained behind. Hahn approached Joliot at the dinner and told him that he believed his wife's results were wrong. The argument was, of course, about the key issue governing the behavior of uranium: Did uranium, when bombarded by neutrons, form heavier elements beyond those known to science, or did it somehow split into smaller entities? This was the biggest scientific problem of the day—and the most portentous from the viewpoint of history.

When Frédéric Joliot returned to Paris and told his wife about the encounter, she was angry. Careful scientist that Irène Joliot-Curie was, she repeated the experiment, obtained exactly the same results, and published them again. In Berlin, Hahn became more determined to prove her wrong. He was furious, muttering curses at Curie while pacing in his lab. He had other worries as well—he was about to lose his partner. Given the ascendancy of Adolf Hitler and the National Socialists with their campaign against the Jews, it was becoming clear that Lise Meitner would not be able to stay much longer in Berlin.

In 1938, when the Nazis annexed her native Austria, new racial laws annulled the citizenship of Austrian Jews. Legally Meitner was now a German subject. The writing on the wall was clear but the paths of egress were blocked. If Meitner were to escape, it would have to be in secret. Hahn was in no danger and had no need or desire to leave Germany; so their team was about to lose its co-leader. As Lise prepared to leave Berlin by train under cover of darkness, Hahn came to their lab to bid her farewell.

Meitner's departure from Berlin would have a strong impact not only on her personal life—she would remain a refugee throughout the war—but on science. No longer in her Berlin lab, she would have to interpret laboratory results, conveyed to her by letter from Hahn in Berlin, and use only theoretical deductions to continue her work on the deep puzzle of uranium behavior. Ultimately, her exile from Germany may have contributed to her missing out on a well-deserved Nobel Prize for work she would accomplish after her departure.

7

THE EVENTS
OF 1938

On July 14, 1938, Italy adopted anti-Jewish laws called the Manifesto della Razza. An imitation of the infamous Nazi Nuremberg laws, these statutes relied on spurious science in order to prove that Jews were of a different race than the supposedly Aryan Italians. They excluded Jews from civil and military service, from much of the economic life of the country, and later from the medical profession. To their credit, many Italians opposed Mussolini's edict and only a few scientists were willing to sign the shameful document.

The Jews of Italy had lived for many generations in harmony with the rest of the population, were in large part assimilated, and enjoyed excellent relations with their Christian neighbors. In parts of southern Italy the Jews were so well integrated that non-Jews hardly knew there were Jews living among them—they didn't seem different in any way—and in some regions there simply were no Jews at all. In one incident, the mayor of a small town in Sicily received a telegram, as did many other mayors, instructing him to enforce the new laws by segregating the Jews. He cabled back to Rome, saying: "Agreed. But what are Jews? Please send a specimen."[1]

Laura Fermi came from a Jewish family, but her children with Enrico were raised Catholic, according to his family's religion, and had been baptized. There was no danger to Enrico and the children—but the same could not be said about Laura. In fact, once the Nazis occupied Italy in September 1943, deportations began, and more than a thousand Roman Jews were sent to Auschwitz in October 1943.[2] Fermi was deeply offended and outraged by his government's actions in 1938, and he did not want to take any chances with his family's safety and well-being. He began to explore employment in other countries, especially in the United States.

The Fermis had visited the United States on a number of occasions over the previous years, and Fermi enjoyed America very much, despite the fact that his English was far from perfect, making lectures difficult to deliver and harder for the students to understand.[3] Ironically, given the July anti-Jewish edicts, it was Laura who preferred to stay in Rome—at least as long as was safely possible—while Fermi himself became eager to leave. In the previous few years, Fermi had been offered many prestigious positions outside Italy. These included the chair of theoretical physics in Zurich vacated by Erwin Schrödinger, and a position at the Institute for Advanced Study at Princeton, where Albert Einstein was. Fermi had turned down all these offers, preferring to stay with his group in Rome. Now in 1938, the situation in Rome looked much bleaker and the prospects for the immediate future even worse. Fermi regretted not accepting any of the fine offers he had had. But the family finally decided it was not too late to make every possible effort to leave. Fermi would write back to the four American institutions that had made him offers to see if any of those academic posts were still available.

By this time, Enrico realized that as "His Excellency Fermi"—a highly respected member of the Italian establishment, albeit a reluctant one who did not agree with the political direction Mussolini's administration was taking—his moves could very well be watched. Italy's international prestige would suffer if a scientist of his standing were to desert to another country. The Fermis, therefore, calculated their moves with great care and secrecy. On a short trip to the Alps, Fermi posted letters to the four American universities saying that his situation had changed and that he would now gladly accept an offer. For added security, in case of censorship and any suspicion that might possibly have arisen, the four letters were posted from mailboxes in four different towns, several miles apart. Fermi got an offer from Columbia University, and the family finalized its secret plans. Earlier, they had scheduled an extended family vacation to Switzerland; they now decided that instead of returning to Rome, they would emigrate to the United States. But an unexpected event complicated the situation.

In the fall of 1938, Fermi was in Copenhagen for a theoretical physics meeting at Bohr's institute. There, Bohr took him aside and whispered that Fermi's name was on the short list for a Nobel Prize in physics that year. There was a possibility of a shared award as well— they might give the award to two, or even three, physicists. Fermi returned to Rome in excitement, with hope and renewed worries about how he and his family could escape from Italy.

On November 10 the phone rang early in the morning, and a bleary-eyed Laura was told: "This is a call from Stockholm, may we arrange to speak with Professor Fermi at 6 o'clock this evening?" The day passed in great anticipation, and when the call came through somewhat later than expected, the tension resolved: Enrico Fermi was to be awarded the only Nobel Prize in physics for 1938 for his work on neutron bombardment and the discovery of radioactive elements beyond uranium. As the chairman of the Nobel Committee for Physics carefully summarized in his presentation speech, Fermi received the award for the "discovery of new radioactive substances belonging to the entire field of the elements and for the discovery, . . . of the selective powers of the slow neutrons."[4] History would show the elements produced by the neutron irradiation were not exactly as Fermi's group had believed.

The Fermis decided they would make their escape to America from Stockholm. Immediately after receiving his Nobel Prize, Fermi, his wife, Laura, and their children would board a ship to New York. They would leave Italy with full pretense of going to Sweden for a short duration—just for the Nobel ceremony—and only close friends and their families would know their true intentions; one of these friends would take care of their apartment to make it seem from the outside that things were normal.

Sweden, a country that would remain neutral throughout the Second World War, would prove to be the first stop on the escape route from the burning continent for many scientists fleeing Hitler's ravages. Ironically, Niels Bohr—who would make every effort to stay in his occupied homeland until hostilities ended, would be forced to flee to Sweden and then to England in 1942—hidden in the hold of a military airplane—and from there to America. In the United States, he and Fermi would take part in the greatest scientific wartime enterprise in history—and one that would forever change our planet. Other members of the once-proud Rome physics group eventually left Italy as the situation in Europe deteriorated. Emilio Segrè went to Berkeley, Franco Rasetti moved to Canada. The days of groundbreaking physics work in the Italian capital were over.

On December 10, 1938, Fermi was awarded the Nobel Prize in Physics in Stockholm. After the ceremonies, he and his family visited Bohr in Copenhagen and from there traveled to England. At Southampton, they boarded the *Franconia* for New York. But while Fermi was speaking about the transuranium elements that could be produced by radiation, very different and completely unexpected results were obtained and analyzed in Berlin.

BACK IN GERMANY, Otto Hahn was determined to prove Irène Curie wrong. He hired Fritz Strassmann, a promising young chemist, to help him. Meitner had just left Berlin late in the evening on July 12 and made it to the Dutch border. With her was her nephew, Otto Frisch (1904–1979), the young physicist from Vienna, who admired her greatly. He had followed his famous aunt to Berlin and helped her in her research. At the border, the Nazi officials left Frisch alone, but they took away Meitner's Austrian passport. She sat petrified while the train remained in German territory at the frontier.

For ten minutes she was in terror. Then an SS officer came into the car and handed it back to Meitner without a word; the train continued and crossed into Holland. But staying there was impossible. Too many refugees were in flight, Dutch borders were closed to escaping Jews, and those who did get in had to continue on to other countries. Even a well-known scientist like Meitner was turned away—she was let in but ordered to leave for a third country. So aunt and nephew continued on to Copenhagen, where both had been invited by Niels Bohr to join his excellent quantum and atomic physics team.

In Berlin, Hahn and Strassmann continued their experiments with uranium radiation and Hahn sent daily letters to Meitner in Copenhagen to apprise her of their progress and ask for her advice on interpreting the results. Hahn felt he was sure to prove Fermi right and Curie wrong. But as fate would have it, the opposite would soon happen.

THE DANISH ROYAL family invested heavily in physics. In addition to Bohr's institute, it paid the salaries of the many physicists who came there for various lengths of time from around the world. They included Erwin Schrödinger, Werner Heisenberg, Paul Dirac, Wolfgang Pauli, and now Otto Frisch.

Frisch carried out experiments aimed at supporting the work of his aunt, Hahn, and now Strassmann. In the United States mass spectrometry studies had now revealed two kinds of uranium: one with weight 238, and a rare isotope with an atomic weight of 235. The existence of this lighter uranium had been discovered in 1935 by Arthur Jeffrey Dempster (1886–1950). A Canadian physicist working at the University of Chicago, Dempster devoted his career to applying mass spectrometry techniques to discover stable isotopes of chemical elements. The German researcher Ernest Niehr in 1938 found through his research that the exact proportion of this U–235 isotope was 0.7%. Frisch also joined research efforts aimed at detecting further properties of both kinds of uranium.

Meitner, meanwhile, felt uncomfortable in Copenhagen, despite the exhilarating presence of so many fellow top researchers. She sorely missed Hahn and their collaboration, and more importantly, she saw her nephew making great strides and did not want in any way to eclipse his achievements. She was a world leader in atomic research, and he was just beginning to gain a foothold. She decided to leave for Sweden.

Meitner had left Germany with only ten Marks in her pocket. She had worn a summer dress and had brought almost no belongings. As soon as Hahn let her departure be known, the Nazis confiscated all her property, including every small item in her apartment, and froze her bank account and pension. Penniless and cold in northern Europe, she had to borrow from friends and fell into despair and depression. Her consolation was the letters sent daily to and received from Otto Hahn. But these often brought bad news about her possessions in Berlin. All her lab reports and her books had been confiscated, and Hahn was forbidden from sending her any of her belongings. She was hoping to move to England, where some members of her family had found refuge, but her plan was thwarted. Like Holland, England too was wary of refugees and severely curtailed their numbers.

Lise Meitner had been invited, although in a lukewarm way, to join the research institute in Stockholm founded by Manne Siegbahn, the most influential physicist in Sweden. But in Stockholm too, she remained depressed and uninterested in what was going on around her. Her only interest was Otto Hahn's correspondence. And then one day in late 1938, Hahn reported strange findings. Bombardment of uranium by neutrons had resulted in the formation of barium—an element similar to calcium, with atomic number 56 and about half the weight of uranium—something, therefore, that was very different from the expected transuranic product Fermi had reported creating in his own lab. This was rather in line with Irène Joliot-Curie's results, so abhorred by Hahn. He asked his long-distance colleague to explicate their findings: How could barium (and lanthanum) be produced by this reaction? Meitner's answer would be the most important development in nuclear physics in the century, and one that should well have earned her a Nobel.

8

CHRISTMAS 1938

Science works in subtle ways. Nature reveals herself slowly, and only to those with open minds, people who can free themselves of accepted beliefs and misconceptions. Boundless imagination also helps. Isaac Newton and Albert Einstein come to mind. Newton perceived an unseen force, gravity, and understood that this force not only made apples fall to the ground but also caused the moon to revolve around the Earth, thus perpetually "falling down" toward us. And Einstein, two centuries later, performed his deep "thought experiments" that helped him deduce the bizarre and marvelous fact that time itself is not constant—as humans had assumed since time immemorial. Who but a mold-breaking genius could imagine such counterintuitive truths?

On Christmas Day 1938 Lise Meitner, far from her homeland and most of her friends and colleagues, on a deserted snowy road outside the village of Kungälv in the backwoods of Sweden, had a moment of revelation that was so revolutionary, so powerful, and so contrary to accepted wisdom that it rivaled those of Einstein and Newton. For while every scientist working on radiation at that time had assumed—without proof and relying on flawed data—that a uranium atom, when attacked by neutrons, could be made to yield heavier, transuranium elements, Meitner saw a different reality. She had a sudden insight that constructed the scientific problem differently, and in doing so she discovered an overarching idea that would literally change the world.

A FEW DAYS before Christmas, she received a letter from Otto Hahn in Berlin. She opened it and read:

> 19.12.38 Monday evening. In the lab.
>   Dear Lise! It is now just 11 PM; at 11:45 Strassmann is coming back so that I can eventually go home. Actually there is something about the "radium isotopes" that is so remarkable that for now we are telling only you. The half-lives of the three isotopes have been determined quite exactly, they can be separated from all elements except barium, all reactions are consistent with radium. Only one is not—unless there are very unusual coincidences: the fractionation doesn't work. Our Ra[dium] isotopes act like Ba[rium].[1]

Hahn was describing his wholly unexpected results using a method of analysis invented by Marie Curie. That method was known as fractional crystallization, or, as Hahn called it in his letter, "fractionation." What Hahn and Fritz Strassmann did was to add bromine (an element similar to chlorine, but even more reactive), slowly and in four steps, to their solution containing the results of a nuclear reaction of neutrons and uranium. At each step, a *fraction* of the liquid mixture crystallized (hence the name fractional crystallization). The pair had been working with barium and radium in the solution because these elements were reactive with bromine at different strengths.

Since radium was more reactive (meaning that it acted faster) than barium, they expected the first step (the first fraction) of bromide crystals that precipitated at the bottom of the mixture to be richer in radium and to contain a smaller amount of barium than the second batch, and to contain a decreasing proportion in each of the remaining two fractionized precipitations. But what they found was very surprising: something that appeared to be barium was present in all four fractions at an equal and very high rate. This was completely unexplainable given the assumptions of the experiment.

The scientists involved in uranium research were convinced that transuranium elements would be formed; they certainly did not expect elements as small (in weight and in number of protons and electrons)

as barium. Hahn thought that there had to be a mistake in his measurements. He ran the experiment again and got the same results. In
frustration, he wrote his letter of December 19 to Meitner, which he
closed with:

> . . . all very complicated experiments! But we must clear it up. Now
> Christmas vacation begins, and tomorrow is the usual Christmas
> party. You can imagine how much I'm looking forward to it, after
> such a long time without you. . . . So please think about whether
> there is any possibility—perhaps a Ba isotope with much higher
> atomic weight than 137? If there is anything you could propose that
> you could publish, then it would still in a way be work by the three
> of us![2]

Hahn's closing sentences demonstrate the depth of the common
conviction that only something larger than uranium could possibly be
produced by the nuclear reaction. Indeed, Enrico Fermi had just received a Nobel Prize in part for supposedly finding larger elements
than uranium. Hahn was so convinced by this reported finding that
he felt that if the result of his own experiment was in fact barium, then
it had to be some freak—an unnaturally large isotope of barium that
nobody had encountered before, a weird animal that was very large in
atomic weight since it had to be bigger than uranium, and the atomic
weight of normal barium was only 137.

But Meitner was a visionary. She could think outside the bounds
of widely accepted scientific knowledge. Meitner placed Hahn's disturbing letter in her pocket and packed her bag to go to the village of
Kungälv, north of the town of Göteborg, across the water from Denmark. She had some friends in this area whom she hoped to see during the holiday.

Hahn's letter had arrived at a difficult time for Meitner. Since her
move to Sweden, she had been even more deeply depressed. Born in
November 1878, she had turned sixty a month earlier, and she felt the
sudden onset of old age. She had spent thirty years at the Kaiser Wilhelm Institute in Berlin, fighting to be seen as a serious scientist, not
just a woman working in a lab. Although now a leader in the field of
atomic physics, she had not made a great achievement, something that
could earn a physicist a Nobel Prize.

Just when her research in Berlin, a center of world science in the pre–World War II years, had started to take off, she was sundered from her group. The rise of Nazism forced her to leave her work, her livelihood, and the only source of happiness and satisfaction in her life. Meitner had worked decades to achieve recognition and the respect due a leading scientist, and just when the effort was beginning to bear fruit, she had had to leave Berlin in the dead of night, with hardly any of her personal belongings—not even adequate clothing to keep her warm in the northern European winter. From a top scientist at a prestigious institution, overnight Meitner became a homeless, stateless pauper: a refugee. And her new country did not want to adopt her. Sweden had never been a country of immigration, and it did not have the facilities to absorb refugees—even if one was a famous physicist.[3]

Her career had hit a dead end in Manne Siegbahn's physics institute in Stockholm, where she had found temporary refuge and partial employment. She had been invited as a favor and in lukewarm tones. Without her lab in Berlin, her experiments, and her trusted colleague of so many years, she felt lost.

In desperation, she wrote to her nephew, Otto Frisch, in Copenhagen, asking if he would come to meet her in Kungälv. Frisch was delighted to comply. He had not seen his aunt in a few months—since her departure from Copenhagen—and he knew from her letters that she was depressed. He would make the two-hour train ride from Copenhagen to the Danish coast, board a ferry for Sweden, and continue by train to Kungälv.

It was an emotional reunion. Frisch arrived late at night, after Meitner had already gone to sleep at her hotel room, and he did not want to wake her up. He was looking forward very much to seeing her; he missed her. He was also eager to describe to her a new experiment on radiation he had been planning at Niels Bohr's high-powered physics institute in Copenhagen, and he wanted his aunt's opinion on it. Since she was one of the world's most experienced scientists in radiation work, her input would be very valuable to him. He slept little that night, and early in the morning, when he knew Meitner would be coming down to breakfast after reading the note he had left under her door the night before, Frisch went to the dining room. In his memoirs he described how absorbed she'd been "by a letter from Hahn and obviously worried by it."[4]

Meitner seemed in a trance. "Barium . . . ," she kept repeating to Frisch, "Barium . . . I don't believe it . . ." What was so bothersome to her was that barium, with an atomic weight of 137, was so much lighter than uranium. "How could you chip off 100 particles [since uranium's atomic weight is 238] from the uranium nucleus?" she asked, looking Frisch intently in the eye, "Barium . . . ?" Frisch sat down with her and they ate their breakfast, deep in contemplation. "Perhaps it's an error," Frisch finally said. "No, no," she replied, "Hahn is too good a chemist for that."[5]

There were reasons for the skepticism. Hahn, Strassmann, the Joliot-Curies, and Fermi were all superb scientists. It was precisely because they were so good that others were reluctant to entertain hypotheses contrary to the logical theory supported by these leaders. The mere notion that a smaller element than uranium could be formed when the uranium nucleus met a stray neutron flew in the face of many solidly accepted physical restrictions—or so the scientists thought. The stumbling block was an assumption about energy.

Only small particles had actually been observed to emanate from nuclear reactions, such as neutrons, electrons, and helium nuclei (alpha particles)—all of them much, much smaller than a barium nucleus. For barium to be a byproduct, the uranium nucleus would have to split. And a calculation had been carried out by a number of physicists showing that the amount of energy required to break a nucleus into large pieces would be enormous. To accept barium as a result of the interaction of uranium with a neutron could be likened to finding a dinosaur hatching from a chicken egg. It was bizarre, unexpected, and seemingly impossible. But because Meitner had more faith in Hahn's abilities as a chemist than in any prevailing physical theory, she could not let go of what she had been told.

Frisch had brought his cross-country skis with him. This trip to the Swedish countryside was an opportunity for outdoor exercise he did not want to miss. So after breakfast he suggested to Meitner that they go outside to explore the woods. Perhaps she could rent skis, he said. "No, no," she insisted, "I can walk as fast as you ski." He laughed and accepted the challenge. They left the hotel and headed west on a trail leading into the lovely woods. And indeed, Meitner did not disappoint him—she could keep up with her nephew's pace for a good distance. Side by side as they progressed, they continued to discuss Hahn's puzzling results.

There were then two theories to describe the atom and the nucleus at its center. One was Ernest Rutherford's theory that viewed the atom as a solid ball, with a much smaller rigid object—the nucleus—inside it. But according to Bohr's newer and more controversial theory, the atom was softer and more flexible, and the nucleus at its center was like a drop of water: its surface tension held it together, but it could change its shape somewhat, the way a raindrop might. Neither theory seemed to permit a splitting of an atom into two pieces, each of them half its original size, when hit by an object as small as a neutron, because the energy required was thought to be immense.

Meitner, walking fast on the trail, and Frisch, gliding alongside her, were discussing these models and how perhaps one conception of the atom might still allow the bizarre behavior reported by Hahn. They stopped and sat down on a fallen tree by the path to rest and admire their quiet, serene surroundings. The snow-covered woods provided a peaceful location in which to discuss a deep mystery. After a few moments of rest, Meitner reached inside her pocket and took out a pencil and some paper. She knew that Rutherford's model would never work—there was no way a solid uranium nucleus could split to form products the size of barium. But Bohr's model held a sliver of hope, she thought—but only if conventional energy calculations (based on the assumed interactions of the forces holding together the nucleus) could be proven wrong. She carefully drew a picture of a raindrop, representing the uranium nucleus under Bohr's assumptions, and then depicted it as shrinking, changing its shape from a sphere to a more elongated, elliptical form; then she drew it expanding again like a spring, and finally splitting into two smaller drops. The picture worked as a visual depiction of the splitting atom. Now she needed to address the energy requirements.

Meitner was not only in excellent physical shape, she also had a phenomenal memory. She recalled the entire, complicated formula used to compute the interactions of the forces holding the uranium nucleus together. She wrote it down by the diagram and began to manipulate the terms of the formula to fit her new assumptions about the atom—once the shape of the nucleus was changed, these forces responded accordingly (something that scientists had not previously considered), and with them the energy required to break the nucleus apart. "The charge of the uranium nucleus, we found," recalled Frisch, "was indeed large enough to overcome the effect of the surface tension almost completely; so the uranium nucleus might indeed resemble a very wobbly, unstable drop, ready to divide itself at the slightest provocation, such as the impact of a single neutron."[6]

Meitner then used her formulas to compute the energy. She calculated that the split of the uranium atom into two smaller atoms, each about half the original size, would result in a loss of mass equivalent to that of one-fifth the mass of a proton. Using Einstein's equation $E=mc^2$, the lost mass would be converted into energy, which she computed to be 200 megaelectron volts (MeV).[7] This energy was just the right amount to allow the two resulting nuclei to exist as separate entities. Suddenly, the pieces of the puzzle fit together perfectly. Stunned by their conclusion, the pair remained seated on the fallen tree for a long time, staring at her calculations in disbelief. The mystery had been solved by her mathematics.

Frisch and Meitner then continued along on the snow together, and she tried to run. Suddenly she stopped and said: "Bohr was right! For this process to work, the atom cannot be hard. It must be like a liquid drop." This conclusion, her drawings that were based on it, and her calculations at last explained the phenomenon that created barium from uranium. And it was in concord with all of Hahn's experimental results. Aunt and nephew then went on to enjoy a Christmas dinner with her friends and decided to write up the results of their discussion as a joint scientific paper. They would do this via long-distance phone calls between Stockholm and Copenhagen—the first lengthy scientific teleconference in history.

TWO DAYS LATER, Frisch traveled back to Copenhagen "in considerable excitement," as he later described it. He was very eager to submit their results for publication, and prior to that, to show them to Bohr, who was about to embark on a voyage to the United States. Bohr could spend only a few moments with Frisch before his ship left port, heading for New York. According to Frisch, when he told Bohr about the new results Lise Meitner and he had derived, "[Bohr] smote his forehead with his hand and exclaimed: 'Oh what idiots we all have been! Oh but this is wonderful! This is just as it must be! Have you and Lise Meitner written a paper about it?' Not yet, I said, but we would at once; and Bohr promised not to talk about it before the paper was out. Then he went off to catch his boat."[8]

Frisch then walked over to another lab in the institute where an American biologist was working. "What do you call a process by which a single cell divides in two?" he asked the American. "Fission," he answered. "Good," said Frisch, "my aunt and I have just discovered *nuclear fission.*" The term Frisch coined that day was used in the joint paper with Meitner that appeared five weeks later in the journal *Nature.* Nuclear fission is the mechanism that would eventually enable the explosion of a nuclear bomb and the production of peaceful electric power from uranium fuel. Meitner, aided by Frisch and using all of her earlier results as well as those of Hahn and Strassmann, had thus derived the most important principle governing the behavior of uranium in extreme circumstances of bombardment by neutrons.

Fission is possible because of an intricate and very delicate dance of particles deep inside the atom—in its nucleus. The nucleus, as we know, contains the positive components of the atom: its protons, which in a neutral atom (rather than a positively or negatively charged ion) are equal in number to the electrons in orbit around the nucleus. The protons "don't like each other"—they exert an electrical repulsion force on one another; although, at short distance, the nuclear strong force dominates. The neutrons that exist in the nuclei of all atoms larger than hydrogen (which has only one proton, hence nothing to repel anything else) mitigate the repulsive effect of like charges. They act as a kind of buffer between and among the protons. These neutrons are, in fact, very similar to protons in many ways, but have become "neutralized"—they carry no repulsive (within the nucleus) positive charge.

The neutrons and protons sit together inside the nucleus and are held together by the strong force—something that Meitner and Frisch viewed as acting like the surface tension of a drop of water. But as the weight and size of the nucleus grows as we approach uranium, the repulsive force inside the nucleus becomes large, and almost intolerable. When we reach the size of the uranium nucleus, the instability attains a level at which the nucleus is so large that the cohesive forces holding it together are weak by comparison.

The uranium nucleus is a kind of wobbly drop, as Lise Meitner imagined it. It is so unstable that when it is hit by yet another neutron (it has so many of them already!), the impact and absorption of the neutron overwhelms the holding force within the uranium nucleus and the nucleus simply breaks in half. When this happens, some small amount of mass is transformed into energy, as dictated by Einstein's famous formula, $E=mc^2$. Since c, the speed of light, is a large number, and squaring it makes it much larger still, we see that a small loss of mass can lead to a (relatively) huge release of energy. The question that scientists would later ask is: Can this reaction—the fission of uranium—be sustained so it will happen to a large number of atoms (or rather, nuclei) of uranium?

WHEN IRÈNE JOLIOT-CURIE read the Meitner, Frisch, Hahn, and Strassmann results in a physics journal, she felt stricken by how badly she had been left behind. Her discovery of lanthanum as a product of the bombardment of uranium by neutrons should have led her to perceive the idea of fission—she had, in a way, observed it in her own experiment. Now this immensely important finding belonged to her rivals—the same group that had tried to discredit her. She regretted that her husband had not collaborated with her on this one experiment in which lanthanum had been detected. Together they too might have found the explanation of fission. Joliot-Curie, generally a highly reserved (and polite) individual, exploded in profanity when she showed her husband the Frisch-Meitner paper: "We've been such dumb assholes!" she screamed.[9]

In an ironic twist of fate, the next step in the discovery process would be made by Enrico Fermi, the man proven wrong when the Meitner and Joliot-Curie teams found smaller elements than uranium and no transuranium results, as he had believed had to happen.

MEITNER REMAINED IN Sweden throughout the war years, and in 1947 was invited to resume her position in Berlin. She refused the offer, despite the appeals of Hahn and Strassmann (a man who would later be honored by the Yad Vashem memorial in Israel for harboring Jews from the Nazis and saving their lives). Meitner agreed to visit Germany, but later emigrated to England. Living in Cambridge, she remained close to her nephew, who would also emigrate to England after the war. She was never awarded a Nobel, while Hahn, Joliot-Curie, and Fermi were all honored in this way for work that was perhaps less deserving. Was it anti-Semitism? Discrimination against women? The fact that she was stateless?

While she never did receive the greatest honor society can bestow on a scientist, most physicists recognized Lise Meitner's great contributions to knowledge. At a physics meeting in the early 1960s, Meitner and Hahn, now both in their eighties, were to be honored. Walking up to the stage together with her old partner, Meitner turned to Hahn and whispered in German: "Walk straight, Hahnschel, or else they will think we are old!"

9

THE HEISENBERG MENACE

Werner Heisenberg (1901–1976) was one of the most important physicists of the twentieth century. But in addition to playing a key role in the development of quantum theory, Heisenberg was also a pivotal figure in the Nazi quest for an atomic bomb.

Enrico Fermi and Werner Heisenberg were the same age—both were born in 1901, one in Rome, Italy, and the other in Würzburg, Germany. Heisenberg's family was deeply rooted in Germany—their genealogical tree stretched back to the early 1700s. His father, August Heisenberg, came from Westphalia, where he taught Greek, Latin, and philology at a gymnasium. In 1909 August Heisenberg was offered a position as professor of Middle Greek and Byzantine history at the University of Munich, and the family moved there.[1] Throughout his life Werner Heisenberg professed a great love for this important city in Bavaria, in the south of Germany. As a boy he attended Munich's Royal Maximilian Gymnasium, where his teachers' reports always commended him for his high performance, talent, and ambition.[2]

As an adolescent, Werner demonstrated strong interest and high ability in mathematics. By age 12 or 13 he had already taken a course in calculus, and would help his older brother with his math courses.[3] He was so enthralled with mathematics that he asked his father to bring him books on the subject from the university's library. August Heisenberg knew nothing about mathematics, so he brought his son any books he found, though he preferred to bring him math books in Latin, his own specialty, thinking that at least his son would learn the language as well as the mathematics.[4]

August Heisenberg served in the First World War, and after being wounded, he returned home in 1916. When his father was well enough to go back to the university, Werner asked him for more books on mathematics. By chance, he brought home Leopold Kronecker's doctoral thesis on number theory, written in Latin. This whetted the boy's interest in number theory, and he wrote a paper on the Pell Equation, an important formulation in the field. Though the journal to which he submitted the article didn't accept it, Werner nonetheless remained interested in mathematics, particularly in number theory. He excelled in mathematics so much that at the age of 16 he tutored in calculus a chemistry graduate student who needed to pass exams in mathematics to qualify for her doctorate.[5]

Although his school was strongly oriented toward the classics, Heisenberg did have a good mathematics teacher. His instructor, recognizing Werner's intelligence, gave him a problem that originated in physics—a problem about the diffraction of light in a vessel of water. Heisenberg tried to solve this problem mathematically using elliptic functions, and this led to a long unpublished paper on the topic. Apparently this work was beyond the level of his teacher, because, Heisenberg recalled, "unfortunately he couldn't say whether it was right or wrong, as he didn't know anything about elliptic functions. But he was a very nice man and he did help me a lot."[6]

Heisenberg was captivated with problems in pure mathematics—including the famous Fermat's Last Theorem, which he tried to prove, unsuccessfully. Then someone gave him the book by Herman Weyl on Einstein's theory of relativity. He studied it and tried to understand the special mathematical tools, called the Lorentz transformations, used in this theory. Yet he didn't think of physics as a profession—it was the mathematics behind the physics that fascinated him. This interest would influence his thinking as a physicist, since it would allow him—like Einstein—to use very complicated mathematics to solve seemingly impossible problems in theoretical physics.

In early 1918 Werner Heisenberg was drafted for auxiliary military service and sent to work on a farm under military-style discipline. He was up at 3:30 AM daily to do physical work all day, which left him exhausted at night. Nonetheless, he thought that the farm service was good experience, since it made him strong. When the war ended in November 1918, he returned to the gymnasium, and after school hours read Kant, Plato, and many other philosophers he admired. In later years, after he had become a leading physicist, he claimed that he first learned about the idea of the atom from Plato's *Timaeus*.[7]

Heisenberg performed exceptionally well on his gymnasium graduation examination in mathematics, and he enrolled at the University of Munich to study pure mathematics. He attended seminars given by the well-known professor Ferdinand Lindemann, who concluded that Heisenberg "knew too much already," and was thus "spoilt for mathematics"—an exceptionally insensitive and silly statement. On hearing of Lindemann's comment, Heisenberg's father (still teaching at the same university) advised his son to take instead the courses offered by the famous theoretical physicist Arnold Sommerfeld, who had expanded on Niels Bohr's first ideas about the atom and had turned them into what is known today as the Bohr-Sommerfeld model of the atom.

Sommerfeld recognized immediately what a bright student he had, so in Heisenberg's very first semester at the university, he gave him an unsolved problem in physics to tackle: the anomalous Zeeman effect. Named after Pieter Zeeman, a Dutch physicist, this effect is the splitting of spectral lines in the presence of a magnetic field. Heisenberg made progress on this problem within two weeks—to the amazement of the professor. Unlike the older physicists who were trained to think "classically," Heisenberg could think about and easily grasp problems in quantum mechanics, one of the two new theoretical approaches to physics (the other being Einstein's relativity). Having started on a problem that was steeped in this "new physics" would help him make his mark as one of the greatest quantum pioneers in the years ahead.

Another student of physics at the university who would also make considerable contributions to quantum theory was Wolfgang Pauli, who became a close friend of Heisenberg. The young Heisenberg did not use his solution to the Zeeman effect as a doctoral thesis because he had been at the university only a short time and felt he had a long way to go before he could qualify to attempt such an major undertaking. He would use it as his lecture a few years later in Göttingen.

In June 1922, the "Bohr Festival," a week-long series of talks by Niels Bohr, took place at the prestigious German university in Göttingen. His lectures attracted physicists from near and far, including Pauli—who was by then a professor in Hamburg—and Sommerfeld and his student Heisenberg. At one of the talks, Heisenberg questioned some of the ideas Bohr presented, initiating a lively discussion. Friedrich Hund, one of the participants, later recalled that everyone's eyes were on the "blond young man from Munich . . . we stared at him in wonder."[8]

But Bohr was not offended or upset—he had found a keen mind, a young person who could really understand quantum theory in depth, at a level beyond that of anyone else in the audience (except for Pauli, as history would show). Bohr invited the young student for a walk up Hain Mountain outside Göttingen. They hiked for hours, and as Heisenberg later put it: "Our discussion, which took us all around the wooded heights of the Hain Mountain, was the first intensive conversation on the fundamental physical and philosophical issues within modern atomic theory that I can remember, and it exerted a decisive influence on the course my life later took."[9] This was also the beginning of a lifelong friendship. Heisenberg was drawn to this founder of atomic theory, and Bohr would later tell his friends that he was surprised to see that the young man seemed to understand "everything." The two would meet again soon.

Heisenberg spent some time working with Max Born in Göttingen while Sommerfeld went to lecture for a semester at the University of Wisconsin. When Sommerfeld returned, Heisenberg finished his doctorate in Munich with a dissertation on turbulence, which was considered brilliant by some, mediocre by others. He then spent another semester with Born in Göttingen, where he presented his lecture on the Zeeman effect, which he had solved in his first semester at the University of Munich.

In September 1924 Heisenberg left Göttingen to go to work with Bohr at his institute in Copenhagen. There Heisenberg began his work on what would be his great contribution to quantum theory—his formulation of matrix mechanics. Completed in July 1925, his derivation allowed for the computation of the stationary states of the atom. Wolfgang Pauli used the new method to solve the states of the hydrogen atom. Along with Einstein's theory of relativity, Quantum mechanics was on its way to becoming a key theory in twentieth-century physics, with the 24-year-old Heisenberg as one of its major exponents.

To crown his achievements, in 1927 Heisenberg would propose a paramount concept of atomic physics—the idea that a particle's momentum and its position cannot *both* be measured with precision, regardless of the accuracy of the apparatus. If one is known with accuracy, the other necessarily entails some uncertainty. This came to be known as the Heisenberg uncertainty principle. Heisenberg claimed that this quality is essential to all behavior in the microworld of atoms, molecules, electrons, protons, neutrons, and similar atomic particles.

Back at the physics department in Göttingen, the only topic of conversation was the new quantum theory. People were obsessed with it—so much so, that the proprietress of the local pub where the physics students and postdoctoral affiliates ate daily once asked them to stop coming in unless they could refrain from their excited conversations about the quantum, as they disturbed other diners. Heisenberg played chess frequently—something his old professor at Munich, the very traditional German academic Sommerfeld, had told him was a waste of his time. As a young man, he would go on ski trips to the Harz Mountains and race, timing himself with the stopwatch the physics department used for experiments. Heisenberg was once clocked skiing at almost 50 miles an hour.[10]

In May 1926 Heisenberg returned to Copenhagen to work with Bohr, fifteen years older and an established figure in physics. It has been said that the complex relationship between Heisenberg and Bohr is central to the history of physics in the twentieth century. In Copenhagen Heisenberg derived the uncertainty principle, and through the many conversations he had with Bohr—often beginning at the institute and ending around midnight over after-dinner drinks at the Bohr home—emerged the "Copenhagen interpretation" of quantum theory. The two men developed the theory, which shatters the old notions of causality, continuity, and the ability of science to know nature with precision. But their conversations were often contentious, difficult, argumentative, and taxing. In February 1927 Bohr left Copenhagen alone and went to ski in Norway. He did not take along his young colleague, as Heisenberg had expected, apparently because he was exhausted by their endless discussions. When he returned in the middle of March, the arguments had come to an end and a new view of the physical world had emerged.

THE PREVIOUS YEAR, 1926, the Austrian physicist Erwin Schrödinger (1887–1961) had developed an alternative approach to quantum theory, using a wave equation named after him. Schrödinger had shown that particles under certain conditions act like waves, and that particle behavior can be analyzed similar to the wave oscillations of a string. His equation could be solved to yield mathematical solutions that agreed with observations. This was a second and equally valid—and often computationally simpler—approach to quantum mechanics. It still competes with Heisenberg's first formulation of quantum mechanics using matrix theory. In September 1927 the quantum physicists, including Bohr, Heisenberg, Fermi, Schrödinger, and Pauli, met to discuss the new physics at a conference near Lake Como, on the Swiss-Italian border, and later most of them continued to Brussels for that year's Solvay Conference, dedicated to quantum physics.

Soon afterward, Heisenberg was given a professorship in theoretical physics at the University of Leipzig. He had to teach there, which he didn't particularly like, but within a few years he had built a physics department that hosted many fine young scientists, including visitors from abroad who stayed for various lengths of time. These included Ettore Majorana, other members of Enrico Fermi's Rome group, and physicists from Copenhagen, Göttingen, and Zurich. Heisenberg won the 1932 Nobel Prize in physics for his work on quantum mechanics. The prize was announced and awarded at the same time as the 1933 Nobel in physics, which was shared between Schrödinger and the English physicist Paul A. M. Dirac (1902–1984), also for their work on quantum theory.

The year 1933 also marked Hitler's rise to power in Germany, and soon the university environment began to change drastically. Hitler launched an attack on Jewish scientists, and within a short time the physics department at Leipzig—as elsewhere in Germany— began to empty of Jewish physicists, who were removed from their jobs and had to move to other countries or remain persecuted and unemployed. Heisenberg refused to take part in a Nazi rally in Leipzig that year and was thus singled out for attacks by pro-Nazi colleagues who considered theoretical physics a "Jewish science." He defended himself, and along with 75 other professors signed a memorandum arguing for the validity of the new physics theories and against the politicization of science. He often visited Bohr in Copenhagen, and his stays there offered him a respite from the increasingly sour situation in Germany. In early 1936 he started giving courses on the theory of the nucleus.

The quest to understand Heisenberg's position on Nazism and the war has fueled intense speculation for decades. We know from his letters and notes that he was upset about events in Germany. On January 11, 1937, he wrote to Bohr on the eve of the latter's departure for a trip to Japan: "The state of the world may have changed very much by the time you are back, and I scarcely dare make plans any longer for more than a few weeks in advance."[11]

That same month, 35-year-old Heisenberg met Elisabeth Schumacher, 13 years his junior, at a musical event in the home of a friend. The two enjoyed making music together, Heisenberg on the piano and Schumacher singing. Within a few months, they married and a year later had twins, followed by a third child.

The verbal attacks on Heisenberg by Nazi elements within the German university system continued. He was accused of consorting with Jews, even though they had been expelled from the universities under the first of Hitler's anti-Jewish laws and the increasing persecutions that followed. Other preposterous accusations, centering on Heisenberg's receipt of the Nobel Prize, were made, including that the Nobel committee was under "Jewish influence" and that he had received the award together with "Einstein's students" Schrödinger and Dirac (that they had studied under Einstein was, of course, untrue, and they were not Jewish, either). The statements in the press reflected many Germans' virulent hatred of the Jews for their purported "continuing influence" on German life and science.

Some of the most venomous attacks against Heisenberg were published in a newspaper controlled by the SS, and he began to feel their ill effect on his career. There was even pressure on him to resign and leave Germany. This situation exemplifies the madness that was gripping the German nation in the wake of the rise of Nazism. By his Aryan looks—fair hair, blue eyes, light complexion—and by his family background, the Nazis could hardly have found anyone more "German" than Heisenberg. And he was a man they should have been extremely proud of, since he brought great honor to Germany with his immense scientific achievements and his Nobel Prize at age 31. And yet, because a relatively large number of theoretical physicists in Germany—before Hitler's purges—were Jewish, Heisenberg was associated with them and thus tainted in the eyes of the Nazis.

Heisenberg decided he could not sit quietly while his name was being smeared and his reputation destroyed. Fortuitously, his grandfather had been friendly with the family of Heinrich Himmler, the head of the SS. Heisenberg wrote to Himmler protesting the published attacks on him. Himmler responded a full year later, on July 21, 1938, writing that he had investigated Heisenberg's case and, taking into consideration family recommendations, had found him innocent of all the accusations; he stated that he would prevent further abuse.[12] Heisenberg's reputation eventually was restored. Perhaps this set of circumstances and events played a role in his adamant determination, as the world slid inexorably into war in the following months, to remain in Germany. He had a young family, he said, and he wanted to stay.

BY THE END of 1938, news of the groundbreaking discovery of fission made by Hahn and Strassmann in Berlin and the explanation and interpretation made by Meitner and Frisch in Scandinavia had reached not only the physics community in Germany but the German military. The possibility of using uranium to build explosives was already known to the Nazis in the spring of 1938. On May 30 of that year, the young physicist Erich Bagge, an assistant to Heisenberg at the University of Leipzig who had given a lecture on deuterium—an isotope of hydrogen used in making heavy water, which plays a role in nuclear physics as a moderator of neutron flow—was approached by two men from the Army Ordnance Research Department.

Bagge was asked if he would agree to speak to them about "nuclear processes" and perhaps accept a job offer. He declined.[13] Soon, the German war department would try for a bigger fish: Heisenberg. Within a year, the Army Ordnance Office in Berlin had recruited Heisenberg—a theoretical physicist with no interest in or experience with experimental issues. They were hoping he could learn to harness the power stored inside the atom's nucleus to produce an atomic bomb.

CHAIN REACTION

The Fermis had a pleasant Atlantic crossing aboard the *Franconia*, although the sea was often rough. They arrived in New York in January 1939 and were initially settled at the King's Crown Hotel, conveniently located near Columbia University. Later they occupied a small apartment near the university, and then they moved to a house in Leonia, New Jersey. This house had a garden that Fermi reportedly didn't like to tend, but he was rumored to have buried a big chunk of his Nobel Prize money in the yard, at least temporarily, before deciding what to do with it. Settling in America was not easy. There were the usual difficulties of language, lifestyle, attitudes, and customs. Given that within less than a year Europe would erupt in the worst conflict in history, these difficulties were minor.

As soon as he arrived in New York, Fermi took up his promised position at Columbia University. Not long after the Fermis left Europe for good, Niels Bohr departed from Copenhagen (right after his brief conversation with Otto Frisch about the Meitner-Frisch results) for a visit to the United States. He arrived in New York on January 16 and, as his ship docked in the harbor, saw Fermi waiting for him on the pier. Fermi promptly grabbed him and took him to his laboratory at Columbia, ignoring any other plans Bohr might have had. That very day, in Copenhagen, Otto Frisch sent the journal *Nature* two papers coauthored with his aunt, Lise Meitner, that explained how the uranium nucleus splits, based on Bohr's model of the atom.

Fermi knew about this extraordinary theoretical discovery, which countered his own conclusion. He was eager to have Bohr join him in his Columbia lab for the start of what could be a most important attempt to observe the process of fission. The experiment began at the Columbia cyclotron (a circular particle accelerator that uses powerful electromagnetic fields to accelerate charged particles, such as electrons or protons, to high speeds and to then crash them into other particles). They bombarded uranium with protons, which are almost identical to neutrons but have a positive electric charge, after accelerating the protons in the magnetic field inside the cyclotron. An oscilloscope (an electrical sensing device with a screen) was attached to the apparatus for the purpose of detecting the fission. Soon after the experiment began, the oscilloscope curve began peaking at regular intervals, roughly every minute, indicating that a uranium atom was undergoing fission. This was hugely exciting: it was the first actual observation of fission.

Unfortunately, by the time these results became evident, only a graduate student, Herbert Anderson, was there to observe it. Anderson recorded careful notes on these crucial experimental results, however. Both Fermi and Bohr had gone to Washington, D.C., to attend a physics meeting. There they met Bohr's student John Archibald Wheeler (1911–2008), an American physicist working at Princeton. The three then traveled together back to New York, where they worked out the documentation of the theory behind the experiment, using the Meitner-Frisch idea to explain how the bombarded uranium nucleus split through the process of fission. Here was the culmination of the immense research effort by so many scientists on both continents. Loyal to his European colleagues, Bohr insisted that his results with Fermi and Wheeler could not be announced until those of Meitner and Frisch were published in *Nature*.

Bohr, Wheeler, and Fermi were joined at Columbia by Leo Szilard (1898–1964), a Hungarian émigré and future Nobel winner who had worked with Meitner. One evening at a dinner table in the Columbia faculty club, they discussed an idea that Bohr had just presented in a talk at Princeton: If the uranium nucleus splits in fission after absorbing a neutron, are *other neutrons* produced in this process?

The answer to this question could lead to the discovery of a chain reaction of uranium fission, fueled by these new neutrons.

There was sound logic behind this hypothesis. Since the uranium atom contains 146 neutrons and 92 protons, once the nucleus is hit by a neutron (produced by a disintegration of another nucleus), it splits in two. But since there are so many neutrons in the uranium nucleus, it stands to reason that when it breaks up, some neutrons would also be violently released through this disintegration.

If this indeed happens, the scientists reasoned, these fission-produced neutrons could in turn hit other uranium nuclei, which would undergo their own fission and release their own neutrons, which would then hit other uranium nuclei, and so on, creating a chain reaction. And, since energy is released through this nuclear reaction (as dictated by Einstein's famous equation), one would conceivably have set up a perpetual-energy machine that could be self-sustaining as long as there were enough uranium atoms to produce more neutrons to hit other uranium atoms. If the process was performed with a large amount of uranium—a *critical mass,* whose level might be determined through complicated calculations—the result of the reaction would be a tremendous explosion. If put together more moderately, large amounts of energy could be produced in a controlled way over time, without an explosion.

It was understood that if a chain reaction was a reality, it could make both nuclear reactors and an atomic bomb realities. The possibility of such a chain reaction was something that Szilard had been contemplating for years, he told his colleagues. Fermi would deal with this key question in the next stage of his research.

LEO SZILARD HAD lofty goals for his life. He believed he was born to be a scientist and to save the world. His scientific career was well on its way when it had to be interrupted because it matured during a tumultuous era. In 1938 he immigrated to the United States, where he pressed the government to work on an atomic bomb before the Germans could make one.[1]

According to his recollections, Szilard's keen sense of justice and morality derived from stories his mother had told him as a child growing up in Budapest. One of them was about his grandfather, a high school student during the Hungarian revolution of 1848. One student was chosen each day to await the teacher's arrival and submit a report on fellow students who had misbehaved in the interim, so they could be punished. On the day Szilard's grandfather's turn came, there was much commotion on the streets because the revolution was going on, and he, along with many other students, left the classroom to join the mobs. Later, he handed his teacher a list of the students who had gone to the streets to join the crowd. It included his own name.[2]

After serving in the Austro-Hungarian army in World War I and being saved from death at the front by contracting the Spanish flu, Szilard went to Berlin to study physics under the famous physicist Max von Laue. He wrote a dissertation on thermodynamics that won Einstein's praise.

In 1933, while Szilard was working in an entry-level academic position in Berlin, Hitler came to power. Unlike some other Jewish academics in Germany, Szilard saw the writing on the wall. He kept his suitcases packed. When things became difficult he escaped to England. Then on September 11, 1933, *Nature* published a speech delivered by Ernest Rutherford to the British Association of the Advancement of Science, in which he said: " . . . to those who look for sources of power in atomic transmutations—such expectations are the merest moonshine."[3]

Szilard found this statement by Rutherford curious, and he tried to think in a contrarian way. He was walking the streets of London one day, and when he stopped at a light, waiting to cross a street, it occurred to him that "if we could find an element which is split by neutrons and which would emit *two* neutrons when it absorbed *one* neutron, such an element, if assembled in sufficiently large mass, could sustain a nuclear chain reaction."[4]

It was within the context of this purely theoretical thinking, which preceded the actual work by Fermi, Meitner, Hahn, and Strassmann, that Szilard had anticipated the idea of a chain reaction. But he did not elaborate on this topic, nor did he publish anything related to it. Although, as he later said, the general idea that energy could be released by such a chain reaction, which could be used to generate power and create a bomb, "became a sort of obsession with me."[5]

He even went as far as to suspect that beryllium might be such an element. If it underwent fission, he reasoned, it might release more neutrons than it absorbed and would thus sustain a chain reaction. Of course, it was beryllium that would be found as the *product* of a nuclear reaction, rather than the progenitor of the process, by the Meitner-Hahn-Strassmann team some years later. At any rate, thinking ahead of his time and being morally cautious—already worried about the possibility of nuclear warfare, according to his reminiscences—Szilard went ahead in 1934 and filed for a patent on a chain reaction, which he assigned to the British Admiralty (British patents no. 440,023, application date March 12, 1934, and no. 630,726, application date June 28, 1934).[6]

Thus, the patent for a chain reaction was filed in England long before the process was proven possible either in theory or in the laboratory.

SOMETIMES IN SCIENCE researchers working in different locations arrive at the same results neck and neck. This was what happened in early 1939, when not only Fermi, Bohr, Wheeler, and Szilard in New York deduced the possibility of a chain reaction of uranium fission, but Irène Joliot-Curie and her husband, Frédéric Joliot, in Paris had the same idea. What worried some scientists was that if Otto Hahn, in Nazi-controlled Germany, also understood this, the Nazis might compel him to use the concept of a chain reaction to produce a bomb.

Leo Szilard left the group in New York and traveled to Princeton, where Einstein was now at the Institute for Advanced Study. It was Einstein's equation $E=mc^2$ that provided the theoretical basis for this release of energy from a chain reaction, and Einstein, who had immigrated to the United States in 1932, was the best-known scientist in the world. Szilard discussed the Columbia experiments with Einstein, who agreed with him that a chain reaction was possible. Szilard then urged Einstein to write his now-famous letter to Franklin Delano Roosevelt, warning the president that the Nazis could obtain such a doomsday weapon, and imploring the president to embark on a project that would allow the U.S. to get one first. Einstein wrote his letter on August 2, 1939. Its effect would be seen two years later.

ACROSS THE ATLANTIC, in England, work on understanding uranium processes was progressing rapidly. In early 1940, Otto Frisch together with Rudolph Peierls at the University of Birmingham started to work on the problem of how much of the lighter isotope of uranium, with weight 235, was required to create an atomic bomb.

Peierls was German, a student of Arnold Sommerfeld, and had obtained a Rockefeller Foundation grant to work with Enrico Fermi in Rome. While there, he had been offered a position in Hamburg that was a stepping stone to a professorship, which made it very attractive. But conditions in Germany were deteriorating just before Hitler's ascent to power, and Peierls decided to go to Britain instead—which at that time did not offer many academic research positions. He got a two-year position at the University of Manchester where a fund had been set up to support German refugees by providing them with fellowships.[7] Some years later, he began to work with Frisch.

The pair of scientists came to a startling conclusion: based on their theoretical calculations, the amount of uranium 235 necessary for a bomb was not measurable in tons, but rather in pounds. While they did not obtain the exact value, their conclusion, referred to as the Frisch-Peierls memorandum, written in April 1940, was the first document that indicated the construction of an atomic bomb was actually feasible. The pair concluded that an atomic bomb should constitute two components whose combined mass would be critical, and hence would explode when brought together, while each separate part would remain subcritical, so the bomb could be handled without exploding before the desired moment. These results led directly in Britain to the creation of an official committee devoted to atomic research. About that time, Niels Bohr sent a telegram to Otto Frisch, which ended in the words "and tell Maud Ray Kent." Frisch and his colleagues assumed this was a coded message, which they unsuccessfully tried to decipher. Because of the connection of the whole project to Bohr's work on the atom, British officials decided to name their atomic committee the Maud Committee. After the war, it was discovered that Maud Ray had been Bohr's children's governess, that in 1940 she lived in Kent, and that Bohr had simply wanted to send her his regards from wartime Copenhagen.[8] The Maud Committee was to be responsible for atomic research in Britain.[9] Late in 1940, wartime scientific cooperation between the United States and Britain intensified, and American physicists were invited to attend Maud meetings.

On December 14, 1940, a team led by Glenn T. Seaborg, working with the 60-inch cyclotron at the University of California at Berkeley, produced the first speck of plutonium 239 by bombardment of uranium with particles inside the cyclotron. The element was named after the planet Pluto (an element called neptunium, after the planet Neptune, had been discovered earlier). With a half-life of 24,100 years, this element is not found in nature, but another isotope of plutonium, with weight 244 and a half-life of 80 million years, does exist in trace amounts in nature.

In the summer of 1941, the Maud Committee, having worked with British and American physicists, issued a report that stated: "We have now reached the conclusion that it will be possible to make an effective uranium bomb . . . likely to lead to decisive results in the war."[10] The report also made a mention of the new radioactive element produced artificially, plutonium, as a possible source of fission for a nuclear bomb. The Americans received the Maud Committee report on October 3. Vannevar Bush, an electrical engineer who was head of the National Defense Research Committee established by President Roosevelt in June 1940, received the report from Britain and decided to take it directly to the President.

These developments taking place on both sides of the Atlantic and reinforced by the Maud report's conclusions led President Roosevelt to convene a meeting in the White House on October 9, 1941. In this meeting, the fateful decision was made: The United States would embark on a major scientific-industrial-military effort to produce an atomic bomb. The first $6,000 allocated went to Fermi's group at Columbia.

Once the Americans made their decision, it remained to the British—who had made serious advances in the theory of nuclear processes—to establish their role in the project. England was being bombed nightly by the German air force and to continue the necessary large-scale research project in that country would have been risky. On July 31, 1942, Churchill made the final decision that Britain would support an atomic bomb effort centered in the United States and controlled by America.

In the United States a special department of the Army was created to control and coordinate the project. Initially, this body was based in New York City and named the Manhattan Engineering District. The effort would later be headquartered at Los Alamos, in the New Mexico desert, but the entire operation of developing an atomic bomb would continue to be called the Manhattan Project. In September 1942, General Leslie Groves was named its commanding officer.[11]

But what were the Americans and British facing? Did the enemy side have a serious on-going effort to produce an atomic bomb? This is one of the most troubling questions about the Second World War, and it has never been fully answered. In several crucial ways, the answer hinges on understanding the wartime behavior of one man: Werner Heisenberg.

Albert Einstein, whose theories of relativity changed the way we think about the universe. From the Library of Congress

For his discovery of spontaneous radioactivity Henri Becquerel shared the Nobel Prize for Physics in 1903. From the Library of Congress

(left) *Sir Ernest Rutherford discovered that the atom has a charged nucleus at its center. From the Library of Congress*

(right) *Danish physicist and Nobel laureate Niels Bohr was a major figure in twentieth century physics. From the U.S. Department of Energy*

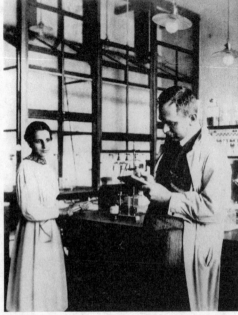

(left) *Lise Meitner and Otto Hahn in their laboratory at the Kaiser Wilhelm Institute in Berlin. From the U.S. Department of Energy*

Marie Curie, fourth from left, with President Warren Harding. From the National Photo Company Collection, Library of Congress

Irène Joliot-Curie found herself in competition with Lise Meitner and Otto Hahn to discover how the uranium atom behaves. From the Library of Congress

Enrico Fermi explaining his theoretical work on the blackboard. From the U.S. Department of Energy

Stagg Field, University of Chicago, below which Enrico Fermi achieved the first nuclear chain reaction. From the U.S. Department of Energy

A sketch of the first nuclear reactor, Chicago Pile 1, which was constructed under the football grandstands at the University of Chicago. From the U.S. Department of Energy

A sketch of Fermi and other scientists gathered next to the Chicago Pile 1 reactor as it produced the world's first chain reaction in 1942. From the U.S. Department of Energy

An experiment with uranium. From the U.S. Department of Energy

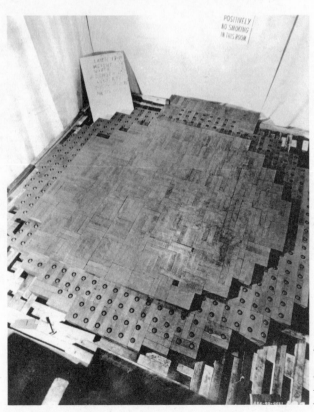

Construction of the Chicago Pile 1 reactor. From the U.S. Department of Energy

Hungarian-born physicist Leo Szilard played a role in nuclear research and in efforts to prevent the use of atomic bombs. From the U.S. Department of Energy

The mushroom cloud from a test explosion of an atomic bomb in the Pacific Ocean. From the Farm Security Administration– Office of War Information Photograph Collection, Library of Congress

A museum display of a model of the Fat Man, the atomic bomb dropped on Nagasaki. From the U.S. Department of Energy

J. Robert Oppenheimer and Leslie Groves at Ground Zero at the Trinity site, September, 1945. From the U.S. Department of Energy

11

THE NAZI
NUCLEAR MACHINE

great opportunity to avoid the frantic race to build an atomic bomb
and perhaps to prevent the death of hundreds of thousands of innocent people and the arms race that followed the war was lost to
humanity because of the stubbornness of Werner Heisenberg.

While the United States and Britain prepared for the possibility of harnessing the power of the atom in a doomsday bomb as a response to the
eventuality that the Nazis might obtain such a device, Heisenberg was tour-
ing the United States. At numerous universities he met with physicists,
some of whom were recent émigrés from Europe. In Berkeley, California, he
met with Robert Oppenheimer; at the University of Chicago and the University of Michigan he spoke and met with other physicists; he met with
Hans Bethe, a German-born physicist who had studied with Fermi in Rome
and had recently immigrated to America; he met with Victor Weisskopf and
Eugene Wigner, both prominent American physicists of European origin;
and he met with Isidor Rabi at Columbia University. And, of course, he met
with Enrico Fermi.

All these men of science pleaded with Heisenberg to remain in America and bring over his family. He could easily have obtained a position with
excellent working conditions at Princeton, Columbia, the University of
Chicago, or any of a number of prestigious American universities. Most of

the best physicists in the world were now in the United States or in Britain—Hitler's purges had produced this brain drain that eventually hurt the Third Reich's scientific efforts. Heisenberg would have had superb colleagues with whom to work and exchange ideas. Any institution would have paid handsomely to have the brilliant Nobel laureate on its faculty, and the Nobel money could have helped Heisenberg settle his family comfortably in America. Besides, it was clear to everyone he met, as it must have been to him, that Europe was on the eve of a terrible war.

And yet, repeatedly, Heisenberg said no, explaining that his family was in Germany, that he was a German, and that his country needed him. He was deaf to the reasoned, logical appeals of his many friends in America. Even after his bitter personal experience with the power of the Nazis and seeing what they'd done to so many of his Jewish colleagues, he refused to leave Germany. He returned home in mid-August aboard the *Europa,* a ship nearly devoid of passengers since no one wanted to travel to a continent about to explode in war.[1]

On September 1, 1939, when Germany invaded Poland, the long-anticipated war started. Since Heisenberg was an army reservist in the Mountain Rifle Brigade and had practiced mountain climbing and target shooting every year with his unit, he expected to be called to the battlefront any day. What he did not know was that his former assistant Erich Bagge had apparently changed his mind, or had been pressured strongly enough, and was now working for the Army Ordnance Research Department as part of a group of scientists recruited to study uranium fission, informally called the "uranium club" (or "uranium society," *Uranverein,* in German, sometimes written *Uranium Verein*), and that this group had also been joined by one of the co-discoverers of fission, Otto Hahn.

The *Uranverein* had been established before the start of the war by the Nazi government in an effort to produce an atomic bomb. The project took place at a dozen different sites scattered throughout Germany, at first mainly in the physics research institute at Berlin-Dahlen. Bagge had asked the director of the Berlin unit to recruit Heisenberg, thinking he was doing the latter a favor as he knew that his professor was in the army reserves and feared he would be called to the front. Bagge met with resistance, however, as there were some anti-Heisenberg feelings due to an old academic dispute, and because the participants saw the uranium research as purely experimental while Heisenberg was known as a theoretical physicist.

But as the military heads of the Ordnance Research Department were pressing scientists to find a way to produce a fission bomb, Otto Hahn continually raised objections. Since he had played such a crucial role in the discovery of fission he was listened to politely. But his new employers did not want to hear what he was saying. Hahn kept bringing up reasons why their effort could not work.

One reason for Hahn's reluctance was a paper he had read by Niels Bohr and John Archibald Wheeler. It revealed the results of experiments indicating that only the rare isotope of uranium, U–235, underwent the fission observed in samples of raw uranium, which included both U–235 and U–238. Although Germany possessed huge reserves of uranium in the Joachimsthal mines, Hahn argued that there was no easy way to separate the very rare U–235 isotope from the mixed ore.

But the authorities did not want to hear "no," and persisted in their pressure on the scientists, who may have thought they had found an easy way to avoid front-line duties by continuing to do their own research in the Berlin facilities run by the army. Bagge again suggested that a theoretician of Heisenberg's caliber might help to overcome their problems, and enable them to build a nuclear weapon. Heisenberg was called up on September 20, and on September 26, 1939, he arrived in Berlin to report for his war duties as a theoretical physicist in the Nazi army's atomic research project.

Eventually Heisenberg was recruited to do work at another facility, in the town of Haigerloch in the Black Forest area of southwest Germany. "*Loch*" in German means "hole," and while this city's founding dates back to the eleventh century, the term *loch* in its name is appropriate for what happened here during the war years. Under a church built on a rock in this picturesque town, German workers and scientists, led by Heisenberg, dug a deep hole in which they proceeded to build a primitive nuclear reactor. Although the reactor had very poor safety features, Heisenberg conducted atomic research here. In addition to uranium, the reactor required heavy water to moderate the nuclear reactions (in heavy water the hydrogen is not the usual hydrogen, but rather a heavier isotope, deuterium, with an additional neutron in its nucleus). Heavy water was supplied by a plant in Norway, which the Allies later successfully destroyed in a series of daring bombing missions both on the ground and from the air. Some German scientists have claimed that the disruption of the supply of heavy water from Norway contributed significantly to the premature ending of Germany's atom bomb project.

In statements they made after Germany's surrender, the German scientists who worked on the Nazi atomic bomb have naturally made every effort to distance themselves from their unsavory contribution to Hitler's war effort. Otto Hahn does not mention his atomic bomb work for Hitler in his biography, *My Life,* published in 1968, although he does discuss his arrest by the Allies and his internment at Farm Hall, a large mansion surrounded by a well-tended English garden in the countryside outside Cambridge.

Much of what we know about the Nazi quest for a bomb comes from secret recordings of conversations among the German scientists interned at Farm Hall, and also from documents discovered by the Alsos Mission in Europe in 1945.

The Alsos Mission was the code name given to a secret Allied unit that, after the Nazi surrender in May 1945. It was charged with locating and detaining the German atomic scientists as well as with collecting all documents that could be found about the Nazi atomic project. The operation was led by the Dutch-born American Jewish physicist Sam Goudsmit, who knew many of the German scientists personally from joint work he had done with them in Europe before the war. The team found and dismantled the Nazi nuclear reactor at Haigerloch, and eventually arrested Heisenberg, who had been hiding in his country house in southern Germany. They also located and detained nine other scientists who had been involved (one of them only peripherally) in Hitler's atomic bomb project.

In addition to Heisenberg, the detainees included Otto Hahn. For a short while these scientists were kept in a chateau in France and then were secretly flown to England, where they were confined to the grounds and spacious mansion of Farm Hall.

It would have been nearly impossible for them to escape from Britain, and guards were placed in the mansion. But given these restrictions, they had the freedom to move around within the estate, and sometimes, when accompanied, they could even leave to attend scientific lectures and other events. But, of course, they had no illusions: They were essentially prisoners of war, kept there "at His Majesty's pleasure," as they were told when they pressed their case to the British officer in charge of them. Apparently there were no international laws that allowed the detention of noncombatant scientists who had nothing to do directly with the military or the political system. The Allies relied on a British law that permits holding any person for a limited, six-month period "at His Majesty's pleasure."

The ten German scientists spent six months in Farm Hall. What they did not know (although one of them suspected but could find no evidence of it) was that Farm Hall was thoroughly bugged by the British intelligence service. Microphones were hidden everywhere, and all conversations among these men were recorded in an effort to find out how close the Nazis had come to obtaining an atomic bomb and what they knew about the subject.

On February 24, 1992, the British government finally gave in to pressure from the world press and the scientific community and declassified the transcripts of the conversations. The scientists had tried to figure out why they were kept there, entertained plans of escaping, and talked about various issues of common interest.

Major T. H. Rittner, the British commanding officer in charge of the detainees at Farm Hall, included in his report about the German scientists summaries of his assessments of the personalities of his ten charges. These evaluations began soon after the scientists were captured in May 1945 and were revised after they all heard about the atomic bomb dropped on Hiroshima on August 6, which was the defining event in their six-month captivity. Rittner's character capsules of the German scientists are condensed below:[2]

Professor Max von Laue.
A shy mild mannered man. He cannot understand the reason for his detention as he professes to have had nothing whatever to do with uranium or the experiments carried out at the Kaiser Wilhelm

Institute. He is rather enjoying the discomfort of the others as he feels he is in no way involved. Appears, from monitored conversations, to be disliked by his colleagues. He has been extremely friendly and is very well disposed to England and America.

Professor Otto Hahn.
A man of the world. He has been the most helpful of the professors. Unpopular with the younger members of the party who consider him dictatorial. Has a very keen sense of humor and is full of common sense. He is definitely friendly disposed to England and America. He has been very shattered by the announcement of the use of the atomic bomb as he feels responsible for the lives of so many people in view of his original discovery. He has taken the fact that Professor Meitner has been credited by the press with the original discovery very well although he points out that she was in fact one of his assistants and had already left Berlin at the time of his discovery.

Professor Walter Gerlach.
Has always been very cheerful and friendly, but from his monitored conversations is open to suspicion because of his connections with the Gestapo. As the man appointed by the German Government to organize the research work on uranium, he considers himself in the position of the defeated general and appeared to be contemplating suicide when the announcement was made.

Professor W. Heisenberg.
Has been very friendly and helpful and is, I believe, genuinely anxious to cooperate with British and American scientists although he has spoken of going over to the Russians. Has been accused by the younger members of the party of trying to keep information on his experiments to himself. He has taken the announcement of the atomic bomb very well indeed.

Professor P. Harteck.
A very charming personality and has never caused any trouble. His one wish is to get on with his work. As he is a bachelor, he is less worried than the others about conditions in Germany. He has taken the announcement of the atomic bomb very philosophically and has put forward a number of theories as to how it has been done.

Professor C. F. von Weizsacker.
Outwardly very friendly and appears to be genuinely cooperative. He has stated, both directly and in monitored conversations, that he was sincerely opposed to the Nazi regime and anxious not to work on an atomic bomb. Told Wirtz that he had no objection to fraternizing with pleasant Englishmen but felt a certain reluctance in doing so "this year when so many of our women and children

have been killed." Being the son of a diplomat he is something of one himself. It is difficult to say whether he is genuinely prepared to work with England and America.

Doctor H. Korsching.
A complete enigma. On the announcement of the use of the atomic bomb he passed remarks upon the lack of courage among his colleagues which nearly drove Gerlach to suicide.

Doctor K. Diebner.
Outwardly friendly but has an unpleasant personality and cannot be trusted. He is disliked by all the others except Bagge . . . is very worried about his future and has told Bagge that he intends to send in a formal request to be reinstated as a civil servant. He hopes we will forget that he was a member of the Nazi Party. He says he only stayed in the party as, if Germany had won the war, only Party members would have been given good jobs.

Doctor K. Wirtz.
A clever egoist. Very friendly on the surface, but cannot be trusted. He will cooperate only if it is made worth his while.

Doctor E. Bagge.
A serious and very hard-working young man. He is completely German and is unlikely to cooperate. His friendship with Diebner lays him open to suspicion.

THESE CHARACTER IMPRESSIONS give us an idea who the scientists involved in Hitler's quest for a bomb were. But what had they really done? More than six decades after the end of the war, the answer to this question is still not known with certainty.

As far as we know, Max von Laue was not involved in the Nazi atomic project. At 66, he was older than the other scientists at Farm Hall, and had done groundbreaking physics decades earlier when he discovered X-ray diffraction—a phenomenon still exploited in crystallography today—for which he received a Nobel Prize. Laue supported Einstein against anti-Jewish attacks on relativity theory, and he generally expressed views opposed Nazism. He was in Berlin during the war, but from whatever records remain, he took no part in the effort to build a bomb.

Otto Hahn's main culpability was that he remained in Berlin during the war and continued his laboratory work on uranium fission—the same research project he had embarked on years earlier with Lise Meitner. It is not clear that any of his wartime work in Berlin was used directly by the Nazis in a significant way, but he was a member of the *Uranverein,* and thus officially involved in Hitler's push to create an atomic bomb. Hahn later claimed that by staying at his lab he was able to continue to employ a number of scientists who might otherwise have been dismissed and would have suffered in the harsh conditions of wartime Germany.

Walter Gerlach was a well-known German physicist who had done pioneering work in quantum theory and co-discovered with Otto Stern the Stern-Gerlach effect while studying the magnetic moment of silver atoms in 1922. The effect they discovered was a deflection of particles in an inhomogeneous magnetic field, and the experimental set-up that produces it is still used today in the study of quantum phenomena. Gerlach became one of the most influential members of the *Uranverein,* studying atomic energy and its generation. In 1943 Gerlach assumed a leadership position over the project.

Paul Harteck was a chemist working in the field of physical chemistry in Berlin, and in 1933, the year Hitler ascended to power in Germany, he went to England to work with Ernest Rutherford. Later, he became the head of the department of physical chemistry at the University of Hamburg. As member of the Nazi *Uranverein,* Harteck was responsible for research on two important problems: the production of heavy water—used in controlling nuclear reactions—and the task of isolating the rare isotope uranium 235 from uranium ore—the key work at the heart of the process that could produce the fissile material for an atomic bomb.

Harteck began working for the army's Ordnance Research Division on uranium and its possible military applications as early as 1937, and two years later he consulted with the Nazi War Department on using fission in the production of an atom bomb. He spent the early war years developing projects aimed at producing heavy water for nuclear research and development, including supervising heavy water production in the plant in occupied Norway. In 1943, Harteck invented a new method of separating uranium 235 from ore using centrifuges—essentially the same process the Iranians are now using six and a half decades later in their ambitious multiyear nuclear project. More than any other scientist, Harteck brought the Nazis dangerously close to their goal. But fortunately, making an atomic bomb required the commitment of larger resources than were available in wartime Germany.

We know that the Nazis did not separate enough uranium 235 using Harteck's centrifuges to actually make a bomb, nor did they make enough plutonium in their Haigerloch facility for the second possible kind of an atomic bomb. Some of the German scientists claimed at various times after the war that they were not really after a bomb—they simply wanted a nuclear reactor for energy.

Kurt Diebner, a member of the Nazi Party, was the organizer of the *Uranverien*. He alerted Nazi officials in 1939 to the theoretically derived possibility of a chain reaction in uranium and its military applications. Because of his zeal for the development of atomic bombs in Germany, Diebner was disliked by some of the other scientists working in Germany at the time. Diebner conducted fission experiments starting in 1942 at Berlin-Dahlen and at universities in other cities in Germany.

Erich Bagge worked with Diebner and invented a process for refining uranium 235 from uranium ore. His method, invented in 1944, after the Nazis were clearly going to lose the war, was more efficient than earlier methods and used a combination of thermal diffusion with electromagnetic fields and centrifugal techniques to separate the fissionable isotope of uranium. Because this development came so late in the war, it never reached its full potential; if it had been actualized earlier, the Nazis would have had a higher probability of achieving their nuclear ambition.

Horst Korsching also worked on the thermal diffusion process used in separating uranium isotopes. He held a position in Berlin and, as a member of the *Uranverein,* performed experiments on uranium enrichment, leading to published work on thermal diffusion. His results were used by Bagge in his work on uranium refinement.

Karl Wirtz, who was a professor at the University of Leipzig before the war, was also recruited for the Nazis' *Uranverein* and was a member of some Nazi organizations. He played a more minor role than some of the other German scientists, working mostly on research on the production of heavy water.

One of the most important scientists who worked for the Nazi atomic project was Heisenberg's close associate Carl Friedrich von Weizsäcker. He was the son of Ernst von Weizsäcker, one of Hitler's top diplomats and one of the foreign office officials convicted of war crimes at Nuremberg for ordering the deportation of Jews to Auschwitz (he was sentenced to prison but released in 1950).

Carl Friedrich von Weizsäcker made an important scientific discovery while working for the *Uranverein*. He deduced that when U–235 underwent fission—as Bohr and Wheeler had shown, it was this isotope that split and released neutrons in the process—some of the neutrons would be absorbed by the heavier isotope U–238. He surmised, in a way that was similar to Fermi's deduction about the production of transuranium elements, that the result of this absorption would be the isotope U–239, which could not be stable and would therefore decay into another element with weight 239. This element, which, as noted earlier, would be experimentally created in a lab at the University of California at Berkeley in 1940, and would be named plutonium, could be used like U–235 in the creation of an atomic bomb. But while the isolation of U–235, which constituted only 0.7% of natural uranium ore (the rest being U–238), would be a very costly enterprise, making the new element in a reactor might prove a feasible route to a Nazi bomb. Weizsäcker reported his results to his superiors in the Army Ordnance office in July 1940.[3]

THE NAZIS CONTINUED to pursue vigorously every possibility for producing an atomic bomb. They pressed Heisenberg, by far the most prominent scientist in the group, for details about the viability of the project. Heisenberg appears to have been ambivalent about this work. He admitted that a bomb was a theoretical possibility—by that time more than a hundred papers had appeared in the scientific literature about fission and about the possibility that a self-sustaining chain reaction of uranium might be realized—but he kept repeating that separating U–235 from a mixture containing mostly U–238 would require many millions of deutschmarks and hundreds of thousands of workers.

The other possibility that had just emerged was to build what the German scientists called an "engine"—a nuclear reactor. This, too, would be costly, but it was possible. Within such a reactor the new element (plutonium, although the Germans did not know about its isolation and naming until after the war) could be made and used in an atomic device. At any rate, scientists could study the nuclear reactions taking place inside the "engine," and perhaps find yet another way to make a bomb whose size was right for loading onto a plane or firing from a large cannon. A small crude reactor was built at the physics department of the University of Leipzig under Heisenberg's supervision. Later, the larger, top secret reactor was built under the church in Haigerloch.

Thus, despite whatever reservations the scientists working for Hitler may have had about their work, through their research conducted at various sites distributed around Germany—in a geographical segmentation that resembled the Manhattan Project in the United States—the Nazis were able to make good progress on many of the projects that would be necessary for making an atomic bomb. This could well have been expected, given that much of the original research on uranium and its properties had been done on German soil and that some of the scientists involved, such as Hahn, remained at their same laboratories, where they could carry out further research.

The Germans understood fission and could produce it in the lab; they understood chain reactions; they knew which isotope of uranium was the right one to use and had developed methods for isotope separation. They knew about using heavy water as a reaction moderator in a nuclear reactor, and they built a reactor at Haigerloch. They also knew that plutonium could be made in a nuclear reactor and used in a bomb as an alternative to uranium 235. All that stood between them and an atomic bomb was time and budget and manpower. Fortunately for the entire world, they never attained those three essentials.

Two important questions are: How deeply involved was Werner Heisenberg in the Nazi atom bomb project? How fully did he support this effort, and where did his true sympathies lie?

12

COPENHAGEN

The now-famous meeting that took place in Copenhagen in 1941 between the two scientific giants Werner Heisenberg and Niels Bohr was a pivotal event in the history of science, but we don't know very much about it. On one side was Heisenberg—the most senior scientist employed by the Nazis and the man whose scientific knowledge could give them the ultimate weapon. On the other was Bohr—an equally brilliant scientist, Heisenberg's former mentor, and a man who could effect change by mediating between scientists working for the Allies and Heisenberg and his collaborators. Some believe this meeting might even have affected the outcome of the war.

As early as 1938 the Nazis were interested in launching an atomic bomb project, but they could not find high-caliber minds because Hitler's anti-Jewish laws had driven so many scientists into exile. The one brilliant star of physics remaining in Germany was Werner Heisenberg (others, such as Otto Hahn and Carl Friedrich von Weizsäcker, were secondary figures in this research). So it was Heisenberg on the Nazi side and a whole collection of superb scientists on the Allied side.

Heisenberg found himself strangely isolated. He used to be part of an international team of physicists—including Einstein, Bohr, Schrödinger, Fermi, and Meitner—who traveled often and freely exchanged ideas. Now he was alone, loyal to a homeland with which he may or may not have agreed. Because the war blocked the flow of information, he did not know what advances were happening in America or England. For this reason, or perhaps because he was working in concert with the Nazis, in late 1941 he visited his former mentor Niels Bohr in Nazi-occupied Denmark.

The details of Heisenberg's wartime visit with Bohr have remained an enduring mystery. We do not know the content of their meetings nor the exact date they occurred: only that the visit took place in the fall of 1941.

Sometime that autumn—according to the American historian Thomas Powers, it was at 6:15 PM, on Monday, September 15, 1941, although accounts vary, and Heisenberg himself thought that it was at the end of October—Werner Heisenberg arrived in Copenhagen on a train from Berlin.[1]

At least one of these meetings is reported to have included Bohr's wife, Margrethe, and some meetings included Heisenberg's associate Carl Friedrich von Weizsäcker. The murkiness of the subject of these meetings and the great controversy about them spilled into popular culture through the very successful play *Copenhagen,* by the English playwright Michael Frayn, which presented speculations about the nature of these talks.

Until recently, most of the information used in building a plausible reconstruction of the content of the Heisenberg-Bohr meeting has been derived from German or German-sympathizing sources. In 1956 Robert Jungk published a book in German about the Nazi push to build an atom bomb during World War II, including recollections by Heisenberg about his 1941 meetings with Bohr. The book was translated into English and published under the title *Brighter Than a Thousand Suns* in 1958.

Jungk took an approach that strongly favored Heisenberg's recollections. He argued that the German scientists knew that if Bohr wanted to, he could immediately establish relations between them and the Allied scientists working on uranium.[2] Heisenberg, according to this view—which became for a time the accepted understanding of events since Bohr did not publicly respond—traveled to Copenhagen to mediate a secret agreement between scientists working on the atom across the war lines, in an effort to avert a nuclear race.

But Bohr, in Heisenberg's view, was not interested in listening or helping. Jungk concluded that Bohr could simply not be approached and "accordingly, when Heisenberg came to see him, he at once assumed an extremely reserved and even chilly attitude towards the pupil who had once been his favorite."[3]

This information was repeated in a book published in England as *The Virus House* (called *The German Atomic Bomb* in the U.S. edition), written by the Holocaust-denying English historian David Irving in 1967.[4] It was Thomas Powers's 1993 book, *Heisenberg's War: The Secret History of the German Bomb*—which drew in part on the work of Irving and Jungk—that Frayn has said first inspired him to write his award-winning play. The story about the Copenhagen meeting as recounted in these sources is essentially the same.

Irving wrote that Heisenberg came to Copenhagen to see Bohr "to ask for his advice on the human issue." According to Irving, Heisenberg was the "high priest" of German theoretical physics who was "going to seek absolution from the Pope" by asking Bohr if he thought that a physicist had "the moral right to work on the problems of atomic bombs in wartime." By this account, Bohr responded by asking if the military exploitation of atomic fission was possible, in Heisenberg's view, and Heisenberg "sadly replied that he now saw that it was."[5]

The idea propagated by all these sources was that the former protégé came to see his master to discuss human and moral issues, seek some kind of absolution for the crime of working on a bomb for the Nazis, and perhaps offer a deal to the Allies that neither side continue to engage in atomic bomb research, in an effort to sabotage their governments' plans. These could be seen perhaps as admirable goals by an idealistic Heisenberg, who was working in difficult wartime conditions for a government whose goals he did not embrace. But history would prove this speculation wrong.

RIGHT AFTER JUNGK'S book appeared, Werner Heisenberg wrote the author a long letter intended to state more precisely his recollections of what had happened in Copenhagen. In the next printing of the book, Jungk felt compelled to include the letter at the end of his chapter dealing with the meeting. In this letter, Heisenberg said that the results of German experiments with uranium and heavy water had brought him and others to the conclusion that it would definitely be possible to build a reactor from uranium and heavy water, to provide energy. In this reactor, a decay product of uranium 239 would be produced with the potential for being "as suitable as uranium 235 as an explosive in atomic bombs." He noted that he and his collaborators did not know any process for obtaining uranium 235 in sufficient quantities using the resources available to them in wartime Germany. They did know, he said, that atom bombs could definitely be made, but apparently he and his co-workers had overestimated the necessary technical expenditure.

Heisenberg claimed that this situation seemed to him and to others in Germany to present a positive opportunity, that "under these circumstances we thought a talk with Bohr would be of value." He recalled that this talk took place during an evening walk in a district near Ny-Carlsberg, and explained that "Being aware that Bohr was under the surveillance of the German political authorities and that his assertions about me would probably be reported to Germany, I tried to conduct this talk in such a way as to preclude putting my life into immediate danger." According to Heisenberg, his talk with Bohr probably started with asking Bohr whether or not it was right for physicists to devote themselves in wartime to uranium research, as there was a possibility that progress could lead to "grave consequences in the technique of war." He claimed that Bohr immediately understood the meaning of his question, as he "realized from his slightly frightened reaction."

Bohr replied, Heisenberg said, with the counter-question: "Do you really think that uranium fission could be utilized for the construction of weapons?" Heisenberg then said that he may have replied: "I know that this is in principle possible, but it would require a terrific technical effort, which, one can only hope, cannot be realized in this war." Heisenberg said that he perceived that Bohr was "shocked by my reply," and surmised that Bohr had assumed "that I had intended to convey to him that Germany had made great progress in the direction of manufacturing atomic weapons." Heisenberg claimed that he tried to "correct this false impression" but that he did not succeed in winning Bohr's complete trust, especially, he said, as he dared only "speak guardedly," because he was afraid that some phrase or another might later be used against him by the Nazis. He concluded by saying that he was "very unhappy about the result of this conversation."[6]

This letter from Heisenberg to Jungk angered Bohr when he read it in 1957. In subsequent statements by Bohr's son Aage and others who purportedly had knowledge about the meeting there have been intimations about how Bohr reacted to the letter and the portrayal of the 1941 meeting.[7] But until recently, no one knew exactly what Bohr's reaction to the Jungk allegations and Heisenberg's letter was.

In 2002, facing immense public interest in the case driven by the success of the *Copenhagen* play, the administrators of the Niels Bohr Archive in Copenhagen felt compelled to act. Bohr's private papers were supposed to remain sealed for 50 years after his death; this would have made them accessible to scholars and the public in 2012. But because of the pressure from many directions to allow an inspection of Bohr's papers for a reaction to Heisenberg's portrayal of their meeting, the directors of the archive relented. They made archive material available ten years before the intended time.

It turned out that Bohr wrote Heisenberg repeatedly, disputing the veracity of Heisenberg's account to Jungk—but never sent him these letters. Bohr was reluctant to confront his former friend and protégé. The warm relationship between the two men had begun to cool even before Germany invaded Denmark. Rumor had reached Bohr that Heisenberg had voiced approval of the Nazi invasion of Poland and had greeted with enthusiasm the German expansion in Europe.[8] By all accounts, their 1941 meeting ended disastrously and the relationship between the two men was never the same again. So in 1956 Bohr did not want to intensify the already bad feelings between the two of them and therefore kept his letters to himself.

These letters can be seen in facsimile, in Bohr's handwriting (or sometimes in that of the person who transcribed them for Bohr—his assistant Klackar, his wife, Margrethe, or his son Aage), in printed form in Danish, and in English translation.[9]

Here are parts of Bohr's first letter in the collection of undelivered letters. It is not dated but according to the Bohr archive was written in 1957:

> Personally, I remember every word of our conversations . . . you and Weizsäcker expressed your definite conviction that Germany would win and that it was therefore quite foolish for us to maintain the hope of a different outcome of the war and to be reticent as regards all German offers of cooperation. I also remember quite clearly our conversation in my room at the Institute, where in vague terms you spoke in a manner that could only give me the firm impression that, under your leadership, everything was being done in Germany to develop atomic weapons and that you said that there was no need to talk about details since you were completely familiar with them and had spent the past two years working more or less exclusively on such preparations . . . If anything in my behavior could be interpreted as shock, it did not derive from such reports but rather from the news, as I had to understand it, that Germany was participating vigorously in a race to be the first with atomic weapons.[10]

Other unsent letters from Bohr to Heisenberg followed, and they expressed even stronger sentiments. The letter marked "Document 11a" in the Niels Bohr archive, again of an uncertain date, reads in part:

> . . . what I am thinking of in particular is the conversation we had in my office at the Institute, during which, because of the subject you raised, I carefully fixed in my mind every word that was uttered. It had to make a very strong impression on me that at the very outset you stated that you felt certain that the war, if it lasted sufficiently long, would be decided with atomic weapons.[11]

Bohr further called Heisenberg's stance "incomprehensible." His damning words contradict everything that Heisenberg said about the nature of his meetings with Bohr, the interpretations of the events in Copenhagen as portrayed in books written on this topic to date, as well as the Frayn play. Bohr raised another issue in his letter: What kind of authorization must Heisenberg have had from the Nazi authorities to visit Bohr in occupied Denmark? The implication is that Heisenberg may have conspired with the Nazi authorities to manipulate Bohr, perhaps in order to derail any bomb project the Allies might have had by pretending to offer an agreement between scientists not to comply with their respective governments.

The previously published accounts failed to even hint at the story revealed by the Bohr letters. The claim that German scientists had actually "sabotaged" Germany's nuclear ambitions has been perpetuated by sustained reliance on David Irving's accounts. This Nazi-sympathizing historian is quoted by Powers seven times in chapter 11 of his book, where Powers deals with the Heisenberg-Bohr meeting.[12]

And Powers recalls that after the war, Heisenberg was quoted as saying, "In the summer of 1939, twelve people might still have been able, by coming to mutual agreement, to prevent the construction of atomic bombs." (But the Copenhagen visit occurred in 1941, after the Nazis had occupied a good part of Europe—surely both parties knew the situation had changed.) Still, Powers states unequivocally that Heisenberg "had just such a 'mutual agreement' in mind when he went to see Bohr," and that "long afterwards Heisenberg told Irving it was 'stupid' of him to seek help and answers from Bohr on matters so difficult."[13]

Heisenberg, of course, cultivated this noble view of himself and his wartime actions with the sympathetic Irving. In the Web site maintained by the Haigerloch Museum in Germany—the museum commemorating Heisenberg's nuclear reactor built in the cave under the church of Haigerloch—one can still find an interview with Heisenberg in which he says: "Very soon after the event there was a meeting in Berlin at which we had to tell the whole story to the officials. Irving has described this in his book, 'The Virus House'. You know the Irving book, don't you? I think it has been very carefully done. He studied all kinds of sources: documents and so on. I think he did it pretty well."[14]

WHILE THE TRANSCRIPTS of recorded conversations from Farm Hall don't shed direct light on the meeting with Bohr, they do cast doubt on Heisenberg's assertion that he and his colleagues attempted to sabotage Hitler's atomic ambitions. A partial transcript of an August 6, 1945, discussion, after the scientists had heard the news about the bombing of Hiroshima, reads:

> HEISENBERG: On the other hand, the whole heavy water business, which I did everything I could to further, cannot produce an explosive.
>
> HARTECK: Not until the engine [nuclear reactor] is running.
>
> HAHN: They seem to have an explosive before making the engine, and now they say "in the future we will build engines."
>
> HARTECK: If it is a fact that an explosive can be produced either by means of the mass spectrograph—we would never have done it as we could never have employed 56,000 workmen . . .
>
> VON WEIZSÄCKER: How many people were working on the V–1 and V–2 rockets?
>
> DIEBNER: Thousands worked on that.
>
> HEISENBERG: We wouldn't have had the moral courage to recommend to the government in the spring of 1942 that they should employ 120,000 men just for building the thing up.[15]

To this, Weizsäcker responded that "all the physicists didn't want to do it, on principle," and that if they had wanted to, they would have succeeded. But I think that Heisenberg's statement in the last paragraph above is most telling: He and the other physicists did not have the *moral courage* to ask the Nazi government for 120,000 workers to make a bomb. What does this mean? It means that they didn't feel it was "moral" to ask the Nazi leadership for these workers, not because what they would be doing (making an atomic bomb) was immoral, but rather because it would have been immoral, in Heisenberg's and his colleagues' view, to take men away from their work—from the front or, perhaps, from the death camps. Even if Heisenberg didn't know about the details of the Nazi horrors, the above remarks are not those of a man who could later claim that he went to Copenhagen to offer the Allies, through Bohr, an agreement not to pursue a bomb because of the immorality of *that* project.

Right after they heard the news of Hiroshima, the German scientists began to work on a joint statement, which they completed and issued on August 8, 1945. They used the opportunity to try to free themselves from any responsibility for the Nazi bomb project. They said that "the fission of the atomic nucleus in uranium was discovered by Hahn and Strassmann in the Kaiser Wilhelm Institute for Chemistry in Berlin in December 1938"—completely ignoring the contributions of Meitner (and Frisch). Then they characterized the achievement as "the result of pure scientific research which had nothing to do with practical uses." The document continued to argue that German scientists were not interested in a bomb because "it did not appear feasible at the time to produce a bomb with the technical possibilities available in Germany," and that their work was purely on producing power from a reactor.

The August 8 statement also tried to create the impression that while the German scientists did possess the knowledge and ability to make an atomic bomb, they did not help the Nazis make one because it would have been too costly and difficult. In other words, maybe the Americans and British truly had solved the riddle of the atomic bomb first, (and some at Farm Hall believed Hiroshima could have been attacked with a very large conventional bomb), but certainly they, the "superior" Germans, could have done it better if only they'd had the means.

In Farm Hall discussions recorded after the news of the second atom bomb, Heisenberg said: "I think that I can imagine all the details of what they have done. The physics of it is, as a matter of fact, very simple, it is an industrial problem." He continued to argue that it would never have been possible to make a bomb in Germany only because of the large scale of the project, and because "all the plants producing heavy water were destroyed by the R.A.F. and that was really why it was not completed. But still, from the scientific side, one knows all the things. The Russians certainly also know it, Kapitza and Landau [Russian physicists]."[16] As with many observers around the world, Heisenberg's eyes were now on Russia as the next major player in the dangerous atomic game.

WHEN I MET Werner Heisenberg while I was a physics and mathematics student at the University of California at Berkeley in 1972, the 71-year-old scientist still looked handsome, walked straight, and stood tall, and his keen blue eyes still had a glint in them. He gave us a brilliant lecture on quantum physics, describing how he came upon the amazing idea of using matrices to describe quantum states—which provided science with the foundation of quantum mechanics and for which he was awarded a Nobel Prize—without even knowing the mathematical techniques for handling matrices. We were all transfixed by his story, and nobody dared ask him what his role in the Nazi atomic bomb project was, and whether or not he had hoped to succeed in giving Hitler a doomsday weapon.

And we will probably never know the answers to these deep and disturbing questions with complete certainty. But Bohr's recently released letters as well as the Farm Hall transcripts have enough in them to cast a deep shadow over Heisenberg's claim of innocence.[17]

13

THE MOMENT
OF TRUTH

n November 1941, a month or two after Heisenberg's visit to Bohr in
Copenhagen, across the Atlantic the U.S. National Academy of Sciences
convened a special committee to review the feasibility of producing an
atomic bomb. The immediate impetus for the establishment of this committee came from the fact that British scientists now agreed with the Americans that a nuclear device was within reach. Furthermore, scientists on both
sides of the Atlantic feared that the Germans were making progress toward
this goal.[1]

Then, on December 7, 1941—"a date which will live in infamy," FDR
called it—the Japanese launched a surprise attack on the U.S. Seventh
Fleet docked in Pearl Harbor, Hawaii, and the United States entered World
War II.

Arthur Compton, a professor at the University of Chicago, was appointed to head the research group that would study the feasibility of eventually constructing a nuclear bomb. The project was named the
Metallurgical Laboratory, and it was centered at the University of Chicago.
Since his arrival in the United States three years earlier, Fermi had been
doing research at Columbia University, studying the production and absorption of neutrons with a small pile of uranium. He was unhappy with
the government's decision to move the project to the University of Chicago

as this required him to commute between New York and Chicago. The new combined scientific group—part of it from Columbia, and part of it from the University of Chicago—was responsible for producing an "intermediate pile" of uranium—that is, one that would contain more uranium, produce more neutrons, and still stay subcritical (less than the amount necessary to set off a chain reaction).

Government policies made Fermi's situation somewhat bizarre. After Pearl Harbor and the declaration of war against the Axis forces by the United States, all Italian citizens living in America were considered enemy aliens and their movements were restricted. Since Fermi was not yet a U.S. citizen, when he needed to travel to Chicago, he had to obtain special permission from the authorities. But the work he did in Chicago was classified as top secret, and he could therefore not reveal to these same authorities the reasons for the requested travel. Many other immigrants found themselves in this paradoxical situation: they had left Axis countries to escape persecution and were now under scrutiny in America.

Fermi moved to Chicago in May 1942, and his family followed soon afterward. (He would remain at the University of Chicago after the war and continue his career in nuclear research.) While his commute under strict government restrictions was over, Fermi had other difficulties as a nuclear scientist who immigrated to America from a now-hostile land. Fermi noticed that all his mail was opened and resealed. This censorship made him furious, and his pride was hurt. He had immigrated to America in good faith and was now helping his new homeland in its war effort—yet he was not fully trusted.

After repeated complaints to the authorities, Fermi was promised that the censorship would stop. But then he found in his mailbox—evidently placed there by mistake—a card to the postal clerk instructing him to open and report the contents of all Fermi's letters. This made Fermi even angrier and he complained to the postal administrator, who claimed he knew nothing about the order but finally offered the explanation that Fermi's letters were being opened by a spy. Fermi found this excuse so stupid that he burst out laughing, and after that it appears the post office left his letters alone.[2]

When Fermi came to Chicago, he entered a period in his life that was characterized by intense secrecy. He went from working in a university physics department to working in a top-secret establishment controlled by the U.S. government—the "Met Lab" as he and others called it. The only thing that Fermi's wife, Laura, knew initially was that the Metallurgical Laboratory employed not a single metallurgist.[3]

Scientists and their family members were to confine social interactions to the group involved in "metallurgical research." They were shown a U.S. government-produced film on secrecy, called *Next of Kin,* which depicted in dramatic terms what could happen if a scientist left a pile of documents unattended where they could be found by a spy. The result might be the destruction of London by the Nazis, the film implied. Fermi, his colleagues, and their family members agreed to comply with strict restrictions on their social lives and their daily activities. The strain was mitigated, somewhat, as the scientific group expanded and their social group grew.[4]

Atomic research in America had reached a crucial milestone. It was clear to all the physicists involved, and had been made known to the military and governmental agencies, that production of a nuclear weapon was very likely. No one had yet produced a significant, self-sustained chain reaction (although some work had been done on this problem at Columbia), which would be the required physical mechanism to enable the functioning of a bomb or a power-generating reactor. It was known, however, that such a chain reaction possibility was likely. As far as a bomb was concerned, the required reaction would have to be fast, and that would be achievable by either using uranium 235 or plutonium. But only small amounts of these isotopes had by this time been produced (in the case of plutonium) or separated (in the case of U–235). It was believed, however, that the amount of material required for an effective bomb would be several kilograms.[5] Estimates of this mysterious and most crucial number, the critical mass, varied on both sides of the Atlantic and across wartime lines. German estimates varied from "a pineapple" to a few tens of kilograms; the British and the Americans thought it was a few kilograms. One of the key pieces of research in the Manhattan Project was to determine the exact amount of fissionable material necessary to make an atomic bomb.

Another important issue was the process used to separate the fissionable material from the raw ore. The Americans considered three techniques to extract U–235: electromagnetic separation, diffusion, and centrifugal methods.[6] A scientific report on all these issues was prepared for the president of the United States. The report also included a discussion of the possibility of using plutonium rather than uranium and separating it out after it was produced within a nuclear reactor. This was to be a major project, with an estimated total cost of over $100 million.[7]

Fermi was chosen to head all nuclear reactor work as he was undoubtedly the greatest expert on neutrons and uranium processes working in the United States. Equally significant, he had expertise in both the theoretical and experimental areas of nuclear physics research. Neutron study was at that time fairly new. In addition to Fermi, Werner Heisenberg and Otto Hahn in Germany, Lise Meitner, then in Sweden, and a handful of other scientists had been working with neutron radiation for only a few years, and the accumulated research on the behavior of these particles was not huge.

Fermi had a unique and almost supernatural "feel" for the behavior of neutrons. For some reason, he was so attuned to them—both fast ones (moving at speeds similar to those of particles in cosmic radiation) and slow ones (those that move at speeds of molecules in a heated substance)—that he could predict what would happen in any given experiment to within statistical error. He always followed his predictions with detailed calculations and these almost always confirmed his intuitions.

Used to directing a team of researchers from his years in Rome, and possessing amazing enthusiasm and a friendly personality, Fermi made the perfect leader in Chicago. Researchers would be busy all around him, reading the results of experiments, making calculations, and calling out numbers. Fermi's eyes would twinkle when he heard what he knew had to be the correct result.[8]

As research on the behavior of neutrons was progressing at the University of Chicago, the government decided to prepare for the construction of various facilities around the country in which to conduct atomic research. The first nuclear reactor, called a "pile," was to be built by Fermi and his colleagues at the Metallurgical Laboratory. Others were to be constructed in Hanford, Washington; at Oak Ridge, Tennessee; and in what would become the center of the Manhattan Project—Los Alamos, New Mexico. The design of these reactors would be based on whatever would be learned from the research done at the University of Chicago by Fermi and his team.

The Fermi team was charged with learning how to sustain a chain reaction of uranium, in which enough neutrons are produced by each uranium atom undergoing fission that the neutrons can trigger the fission of other uranium nuclei they encounter. If too few neutrons are produced in each single-nucleus fission, not enough neutrons will remain for the process to sustain itself. In such a case, after the nuclei that are hit undergo fission and the newly released neutrons are lost, the reaction will stop. So the purpose of the Chicago group was to find out whether enough uranium could be placed in one tight location for the reaction to sustain itself for at least some time.

Construction of the world's first nuclear reactor, the Chicago Pile, began in October 1942. Scientists under Fermi worked 24 hours a day, in two shifts, building an elliptically shaped containment area with a radius of 3 meters and 9 centimeters (about 10 feet), housed in the underground hard-racquet courts under the bleachers of Stagg Field, the football stadium of the University of Chicago. A wide and thick frame made of wood supported the structure. There was a large rubber balloon enclosing the reactor. Fermi had feared that nitrogen in the air might absorb neutrons needed for the nuclear reaction, and the purpose of the balloon was to allow for the removal of all air from the reactor pile, so that all that remained inside was uranium and the moderating materials. As it turned out, the balloon was unnecessary and was subsequently removed.

The reactor's contents consisted primarily of lumps of highly purified natural uranium, uranium oxide, weighing a total of 6 tons. The purest uranium was placed inside the heart of the reactor, and the more mixed matter in the periphery. This way, the neutrons produced in the center of the pile had the highest chance of encountering uranium nuclei to split.

The uranium was buffered by graphite bricks designed to absorb stray excess neutrons and keep the process under control—if these were not there, the whole pile might explode in a nuclear blast once enough uranium was added to the pile, or heat would be produced to such a degree that the whole setup would melt down, as indeed happened 44 years later in Chernobyl (and, to a lesser degree, at the Three Mile Island power plant in Pennsylvania in 1979).

There were also cadmium strips inserted into this pile of layered uranium and graphite. Cadmium is a strong absorber of neutrons, and it was used to keep the neutron emissions under control. Cadmium rods could be lowered or raised, so that their level within the pile was variable, and this provided the scientists with manual control of the emission process. Fermi performed his calculations before each layer was laid onto the growing pile of material. The cadmium rods were attached to wooden handles that could be moved in or out of the pile without exposure to excessive radiation. The handles attached to the cadmium rods were locked in position every night and the keys were in the sole possession of Fermi's two assistants, Herbert Anderson and Walter Zinn. Every day, the scientists measured the amount of radiation—specifically, the neutron flux within the pile—and, after completing calculations, made a decision on whether to raise the rods, thereby increasing the number of neutrons in motion inside the pile. This was essentially a trial-and-error operation: After all the theory had been worked out, it really came down to raw experimentation.

A pile of uranium (or any other fissionable material) goes critical at the point at which the number of neutrons in the mass grows *exponentially*. At first, the number of neutrons flying around inside the pile increases in a slow, linear fashion as the rods that moderate the reaction (by absorbing neutrons) are raised. Then, there comes a point at which the number starts to grow dramatically. That is when the chain reaction becomes self-sustaining. If the cadmium rods are raised abruptly beyond that point, the chain reaction can become violent—because it is too fast, with many uranium atoms splitting all at once—

and the immense amount of heat and other energy will cause an explosion or a meltdown. The trick is to keep raising the rods slowly, day after day, and measure the amount of radiation continuously to determine the critical point.

Fermi was meticulous in personally overseeing all aspects of this experiment. From the prior work of his team at Columbia and given the results of his colleagues in Europe, he was convinced that a chain reaction was inevitable. But could he and his team put together enough uranium in the racquet courts to actually produce a chain reaction—and yet one that was not too big, or else Chicago might disappear in a nuclear explosion?

As the experiment progressed, it became clear to Fermi, whose calculation ability was legendary, that there was a high probability that the critical moment would arrive on the night of December 1. Fermi asked Anderson, whose turn it was to control the reactor during the night, not to let the pile go critical, but to wait for him. Once morning came, it was apparent to Fermi that the pile was ready to become critical. The rods were to be raised very slowly and very carefully now. This was going to be the most dangerous experiment in history.

On the morning of December 2, 1942, Anderson met Fermi at the control area above the reactor and showed him how far the experiment had progressed. They slowly raised the rods higher and higher—a millimeter at a time—taking measurements and making careful calculations. Fermi used his slide rule and paper and pencil to make quick computations of neutron flow. They were very close to criticality, but not quite there. The excitement was palpable, and the forty scientists who worked on the Metallurgical Laboratory project gathered around the reactor, everyone waiting with bated breath to see what would happen.

Fermi was a careful man. He had prepared a number of emergency measures to be taken in case things should go wrong, in case the reactor overheated or started to melt or was about to explode. If the neutron flow became too severe, Fermi had special cadmium rods added to the reactor top, tied with strings above holes in the reactor vessel. Should things somehow go wrong, the strings would be immediately cut, the rods would fall into the shafts below, and the reaction would slow down precipitously, averting disaster. As a secondary emergency measure, Fermi had a number of workers standing by with buckets of cadmium salt solution, ready to pour it down on the reactor and thus shut down the neutron flow.

On the big day the reaction proceeded in an orderly way. The flux of neutrons and splitting atoms was continuing to grow toward criticality without a hitch. Yet, at lunchtime, they were still raising the rods, still waiting. Fermi was never one to miss lunch—not even in the middle of the most important experiment of his career.[9]

At 2:20 PM, the big moment finally arrived. Fermi and his closest colleagues gathered around the devices that measured the intensity of neutron radiation. In one of the most dramatic moments in the history of science, the cadmium rods were brought up the last few millimeters. Then the clicking of the radiation counters and the patterns on the oscilloscopes indicated clearly that a chain reaction was occurring below the ground. No explosion took place. Fermi took a deep breath, and a wide smile spread over his face. The scientists, technicians, and workers burst into spontaneous applause.

The power generated by the Chicago Pile was minuscule—less than half a watt (a small light bulb uses sixty watts of electricity). To minimize the production of dangerous radioactivity, the chain reaction was allowed to continue for only 28 minutes. Once the reactor was shut down, everyone went out to celebrate the successful completion of the experiment. Eugene Wigner, a future Nobel laureate, opened a bottle of Chianti for the occasion. The scientists at this historic event sent a coded message about the success of the mission to one of the directors of the government's nuclear effort. It said: "The Italian navigator has just landed in the new world."[10]

Fermi had produced the first uranium chain reaction. Nuclear fission was thus shown definitively to hold the potential for creating immense amounts of energy from the spontaneous breaking apart of large numbers of uranium atoms and the attendant release of neutrons. Science had made a definitive breakthrough that day under the football stadium in Chicago.

ACROSS THE ATLANTIC, a fateful decision was being made. The Germans were unaware of American and British preparations for making an atomic bomb, and Hitler decided to concentrate his resources on developing rockets to attack London, instead of continuing with the enormous financial support necessary for the Nazis' own atomic project.

This alternative project to build the infamous V–2 rocket that would wreak havoc on Britain was called Peenemünde, and was headed by General Walter Dornberger. Its technical director was the German rocket scientist Werner von Braun, who later came to America and played an essential role in early U.S. space technology development. [11]

Thus, though the Allies were unaware of this fact, they no longer had a serious contender in the race to the bomb. Although the work of Heisenberg and his collaborators at Haigerloch did continue, they just didn't enjoy the extensive financial support of the Manhattan Project.

The scientists in Chicago still had much more to do. A reactor is not a bomb. A reactor is a pile of uranium undergoing a controlled chain reaction. Building a bomb is a far more complicated operation—one that requires an extraordinary effort.

14

BUILDING
THE BOMB

With the reality of a chain reaction firmly established, science was ready to provide an answer to Hitler's challenge to global freedom. Yet the project of harnessing the power of science in an effort to defeat a dreaded enemy would eventually also bring us the horror and cruelty of atomic weapons.

The possibility of nuclear warfare became a reality through the Manhattan Project, the U.S. large-scale, scientific-technical-industrial effort to create the world's first atomic bombs. J. Robert Oppenheimer (1904–1967), a physics professor at the University of California at Berkeley, was selected by the government to direct the research laboratory. A Harvard graduate in chemistry, Oppenheimer had done research in Cambridge, England, and received his doctorate in physics from Göttingen, in Germany.

In June 1942, Oppenheimer held a meeting at Berkeley with several key scientists, including the noted physicists Edward Teller (1908–2003) and Hans Bethe (1906–2005), to discuss the theory and technical aspects of creating an atomic bomb, and to plan work on its production.

The first idea discussed was Teller's suggestion to create a thermonuclear explosion using deuterium. It was believed that the chain reaction would cause shock waves and a release of energy that was quantitatively estimated using extrapolation from conventional devices. Other ideas were discussed as well, including the results of theoretical work done in Britain, mostly that of Otto Frisch and Rudolph Peierls and related work coordinated by the Maud Committee.

It soon became clear that the effort required was enormous and this necessitated forming a consortium of organizations that would perform the needed research. The consortium included the University of Chicago, where the original chain reaction work had been done, the Carnegie Institution's Department of Terrestrial Magnetism, the University of Wisconsin, the University of Minnesota, Stanford, Purdue, Cornell, and the University of California at Berkeley. A site had to be chosen as the central location for the Manhattan Project, and it needed to meet certain military requirements for isolation, logistics, and security. Oppenheimer selected the Los Alamos Ranch School on the Pajarito plateau in the desert of New Mexico, about 40 miles northwest of Santa Fe. This location would be code named the "Project Y Site."[1]

A number of scientific questions had to be answered by the Manhattan Project before the large-scale industrial processes that would lead to the construction of one or more atomic bombs could begin.

Scientists had determined that uranium 235 would be required, but the average number of neutrons produced from the fission of a single uranium 235 atom was not known, although some inaccurate estimates had been made. A definitive number was a crucial parameter to determine if a nuclear explosion was to become a possibility. Plutonium offered an alternative solution, but the number of emitted neutrons for plutonium was not known either. Therefore, one of the main tasks of the theory and experimentation division of the Manhattan Project was a scientific determination of this number for both uranium 235 and plutonium.

A second key scientific goal was to determine the fission spectrum—the energy range of neutrons produced by chain reactions of uranium and plutonium. This work was related to the enrichment of uranium. Uranium enrichment work was already underway at the University of Minnesota, and neutron energy ranges had been studied at the University of Wisconsin and by the group headed by Emilio Segrè at Berkeley. Chemistry and metallurgy research had to be conducted in order to make the physics feasible. The chemistry of uranium had to be better understood and its metallurgical properties learned.

Finally, a way had to be found to put together a critical mass of uranium or plutonium in a device that would create an explosion at a given moment.[2] Here, the Ordnance Division of the U.S. military entered the project. Although the immense complexity of the operation required work in many separate locations throughout the United States, the Los Alamos site was its nerve center.

In the Manhattan Project, the nation's brightest scientific minds were marshaled under Oppenheimer's direction to solve the many remaining scientific and technical problems to create a working atomic bomb. It required enormous effort and a huge amount of money at a time when the country was engaged in a two-front war and the demand for military materials was at its highest. The Manhattan Project ultimately employed 130,000 people and cost $2 billion ($27 billion in today's dollars).

LOS ALAMOS HAD a very complicated command structure. The military ran the project, and it was headed by General Lesley Groves (1896–1970). But the scientists working there reported to the chief scientist, J. Robert Oppenheimer.

The physicist Norman F. Ramsey was the head of the Delivery Group, responsible for converting an experimental device into a usable atomic bomb. Ramsey stated "that if things looked to be going badly at Los Alamos, I had an obligation to report [it] to the Secretary of War. In actual practice this proved not to be necessary, though Oppie [Oppenheimer] did choose to use it as a slight threat to [General Lesley] Groves when something that we really felt was very badly needed at Los Alamos was threatened not to be done."[3] According to Ramsey, "Groves really had pretty much his own show though he was an army engineer."

The atomic bomb project was a unique enterprise—there had never been anything like it in the entire history of the U.S. Army, or for that matter any army ever. It was set up as an individual project, rather than an undertaking whose control was distributed among several army services. The usual procedure in an army development enterprise is that the ordnance division does the ordnance part of the mission, the transportation corps does its part, the engineering corps does the engineering part, and then everything is put together. In setting up this radically different atomic project, President Roosevelt and Secretary of War Henry L. Stimson realized that the normal distribution of tasks would have resulted in a slow process, and time was critical. The Manhattan Project was organized exclusively as a single, whole entity for which all decisions were made by the "Manhattan Engineering District," headed by Groves. According to Ramsey, the Manhattan Engineering District reported "pretty much directly to the Secretary of War. It was, in many respects, outside of the army."[4]

The atom bomb was the project to end all projects. It was military in its nature, but in its status it was beyond war, beyond armies, beyond the scientists themselves. Whoever was first to obtain the atomic bomb would win the global conflict, no matter what the enemy did, no matter how advanced its forces were, no matter how many countries were under its control, no matter how large its conventional forces might be. General Groves could get whatever resources he needed for this highest priority and utterly secret project.

The Ordnance Division had a specially assigned liaison officer just for the Manhattan Project, and Groves would call him and get whatever he wanted: airplanes, bombs for practice runs, trucks, trains, and all sorts of equipment. The building of the atom bomb required skills and equipment, expertise and machinery of so many different kinds that it was indeed the most complex military production project in history. The people who worked on it were extremely dedicated—they worked under immense time and productivity constraints, and they felt that a lot depended on them. The participants shared a sense of purpose, working hard around the clock to beat the Nazis to the ultimate weapon.

The relationship between Groves and Oppenheimer was very complex as well. There were no intermediaries in the chain of command at the base—there were scientists and there were military people. The scientists were Oppenheimer's people, while the military personnel were commanded by Groves. And Oppenheimer reported to Groves.[5]

The difficult issue for the scientists was getting used to military structure in their daily lives. They had been used to doing pretty much whatever they wanted when they worked at universities and research institutions, but they were now working on a top-secret military project. They now had to report to authority and work under tight constraints in absolute secrecy.

The scientists (and even the citizens living in the area) were forbidden to tell anyone that they lived in Los Alamos. In case of emergency, they were allowed to say that they lived in New Mexico but warned to add no details to that. And they were not permitted to say that they worked in laboratories or to give anyone the names of others who worked with them. And, of course, they were strictly forbidden from mentioning that they were chemists or physicists.[6]

The military presence was evident everywhere at Los Alamos: a military post with its security officers everywhere, and guards and soldiers of various responsibilities. There were the usual benefits of an army base—food was more plentiful, especially high-quality items such as meat. And most foods and other goods and services were subsidized by the government and hence were cheaper than elsewhere. But there were disadvantages too. The scientists, their spouses, and children did not like the uniform look of the houses. Similar to barracks, they were painted in the same drab, nondescript military color and their architecture was standard and unattractive. Military rules extended into every aspect of life, including how the furnaces in the houses should be operated, and this meant a general overheating of the homes. In addition, the Los Alamos scientists often detonated conventional bombs in order to test how the bombs exploded under various conditions of weather and air pressure, and to see how these conditions affected their trajectories, explosion patterns, and debris. Los Alamos became known as the "City of Fire." The bombs caused fires in the area so frequently that the military authorities decided to intentionally burn large tracts of forest around the base to minimize the accidental fires.

Military censorship was a major nuisance for the scientists and their families as well. According to Ramsey, every piece of mail sent by anyone at Los Alamos was thoroughly censored. Apparently this irritated most of the scientists and especially their spouses, some of whom enjoyed taking photographs and needed to have them developed. "If you took any pictures they had to be sent to a place where essentially they were developed by the censors and the censors did a bad job of development. As a result, a fair number of us, who have never before or since developed our own pictures, had to undertake doing this at Los Alamos in the hope of not having the pictures just completely ruined."[7]

There was a lot of the usual military red tape and the entrenched military attitude overriding the entire process in which normally free-thinking scientists worked. These conditions presented special problems and challenges. The compartmentalization of the whole Manhattan District project caused serious loss of time, because people working in one area were not permitted any information about those working in other areas—how far they had progressed on the projects and what their needs and special problems were. Thus, scientists working on one aspect of the Manhattan Project, who might have been in a position to help others working on a separate aspect, could not do so.

Ramsey recalled one example where less secrecy within the project and a lesser degree of internal compartmentalization might have resulted in a great savings of time. This had to do with the key problem of the project: isotope separation—obtaining the needed uranium 235 from uranium ore. It had been known to a number of scientists outside Los Alamos that there was a different process of refining uranium, which was ineffective if one started with ordinary uranium ore and tried to separate large quantities of uranium 235 from it, but was more efficient than the diffusion plant method (the one used in the Manhattan Project) once the scientists had reached a level of refinement of about 50% uranium 235 in the mixture. The most efficient process would have been to use the ordinary diffusion process in the early stages, and the alternative process in the later stages.[8] One can only speculate if such time-saving steps might have helped complete the project before the Nazi surrender and thus might have saved more lives, if the Germans had understood that the U.S. possessed the ultimate weapon.

Ramsey argued that one visit from scientists from one plant to another plant would have sufficed for them to realize the greatly increased efficiency that would have been possible. But rigid military rules made a process that could have been speedier blindly inefficient.

In addition, Ramsey recalled, General Groves insisted on commanding every single aspect of the entire operation. Oppenheimer realized that Groves "tended in a certain sense not to be completely frank with each of the heads of the laboratory. Some of it was even for a useful purpose in the long run. Generally, Groves would deliberately give Los Alamos excessively optimistic reports as to what was being done at Oak Ridge." For this reason too, Groves didn't let people from Los Alamos go to Oak Ridge, as they might have found out how slow the other laboratory was on its production schedules. Likewise, Groves used to give the people at Oak Ridge "excessively optimistic reports as to how things were going at Los Alamos, with dominantly, I think, the laudable reason that in this fashion he could make both groups really work hard, since each group would think it was a bottleneck and therefore things would get done faster."[9]

It irked the scientists that their commander was depriving them of important information. In time, the scientists learned that the information they were getting was not entirely honest and complete.[10]

While Groves had certain good qualities as leader of this immensely complex operation and could, for example, procure planes whenever the scientists wanted to drop test bombs on short notice—at a time when every bomber was of crucial importance in battle—his shortcomings were a serious problem for the scientists. "His principal evil," complained Ramsey, "was that he could be rather irritating to the scientists that worked under him. This would have been very bad, I think, under peacetime conditions; under wartime conditions it didn't work out too badly, though I think there aren't many of the scientists who thought highly of him."[11]

THE SCIENTISTS OF the Manhattan Project started working on a bomb design that they called the "gun assembly design." It consisted of a long tubular gun, which would have required major modifications of the B–29 Superfortress bomber that was to carry the bomb to its target. That bomber had two bomb bays, one in front of the other, so the "gun" would have had to stretch from the rear bomb bay to the front bomb bay. Plans were being made to design and build a variant of the B–29 Superfortress that would have one long bomb bay instead of two.

But as the work progressed, the scientists designing the gun assembly bomb realized there was an alternative. The gun design had one subcritical component placed in the "muzzle" end of the gun, separated by the length of the gun from the other subcritical piece. Once the gun was fired, the back piece was blown toward the muzzle piece, becoming one chunk of uranium 235 with a total mass that was supercritical—that is, it weighed more than was necessary to create a chain reaction—and hence the atomic bomb exploded. But the mathematician John von Neumann showed the scientists a second possible design. If one piece of fissionable material that was subcritical could be imploded, meaning collapsed onto itself through the use of carefully laid out explosives placed all around it, then it could become critical and explode.[12]

By early 1945, the nuclear plant at Hanford, Washington, was beginning to produce plutonium of good enough quality for use in a bomb, and the nuclear facility at Oak Ridge, Tennessee, was producing high-purity uranium 235 in sizable quantities. Both facilities sent their products to Los Alamos, where the two different materials would be used as components for bombs of two distinct types: uranium 235 for the gun-design bomb, and plutonium for the implosion-design bomb.[13]

The work at Los Alamos now proceeded along two separate lines: one to produce a gun-design bomb, the other to produce an implosion bomb. The two designs were given code names so that the scientists could communicate with the air force, which was building and modifying the B–29 airplanes able to carry the two different types of atomic weapons, with their different shapes and different characteristics.

The scientists chose code names that, they hoped, would confuse any eavesdropper into thinking that their communications were about preparations for flying the American president and the British prime minister to some meeting. As Ramsey recalled, "it was decided that one of the bombs would be called the tall man, hopefully making people think that it was Mr. Roosevelt, and the other would be the fat man, for Mr. Churchill, and these were the code names used in describing the two [bombs] locally."[14]

And the terms "fat man" and "tall man" then became standard terminology in the Air Force internally.

The group of scientists who worked on designing the tall man found a way to make the bomb shorter. Through their research they discovered that the two subcritical parts of the bomb did not need to be separated by a very long distance. They improved the gun-design bomb to the point that the tubular section could be short enough to fit within a single bomb bay of a Superfortress. This was a major advance as it reduced greatly the extent of the modifications needed in the B–29 bomber—only small changes were now necessary.

But the term "tall man" was no longer descriptive. So the scientists of Los Alamos gave the gun-design bomb a new code name: Little Boy. And the two atomic bombs that were being developed in the Manhattan Project were now referred to as Fat Man and Little Boy, abbreviated as the FM model and the LB model.[15] More than 50 bombs of each of the two kinds, FM and LB—in addition to several dozen of the earlier, now-obsolete Tall Man design—were produced so that their characteristics could be studied. Some were just empty shells without the nuclear material and others contained conventional explosives. They were dropped from airplanes at various altitudes and their ballistic characteristics were studied.

When the two bomb designs were complete and had been tested in flight, and when enough nuclear material to create more than one fission bomb had been produced at Hanford and Oak Ridge, the scientists of Los Alamos were ready for a test. At the same time, some of the scientists were getting ready to go to Tinian Island, in the Marianas chain of the South Pacific, from where the missions to attack Japan with nuclear weapons would take off.

Tinian had been chosen as the base for the atomic attacks on Japan in February 1945, months before any atomic bomb was ready. In a now-declassified top secret memo dated February 24, Commander F. L. Ashworth of the U.S. Navy informed General Leslie Groves about the decision to base the 509th Composite Group, the unit that would be charged with carrying out the atomic attacks on Japan, on Tinian Island rather than on Guam or another U.S.-held territory. He wrote:

> The island of Tinian is 125 miles north of Guam and approximately 1450 miles southeast of Tokyo. It is about 10 miles long and 3 miles wide and has gently rolling terrain. Prior to our occupation, approximately 95% of its area was planted in sugar cane. Tinian has two major airfields, North Field and West Field, the larger being North Field, with four 8500 foot landing strips directed into the prevailing easterly winds . . . It is anticipated that early decisions will be reached and transmitted to the forward area so that construction may be started and the island of Tinian made ready for the basing of the 509th Composite Group upon its arrival during the month of June 1945.[16]

According to another top secret memo (declassified in 1974) sent to the commanding General of the XXI Bomber Command in San Francisco on May 29, 1945, the 509th Composite Group had been organized and activated in December 1944, "by the direction of General Arnold, for the specific purpose of delivering certain special bombs when these bombs become available. It is now anticipated that the first of these bombs will be available for delivery in August 1945."[17]

Once the bomb was ready, the Los Alamos group divided to work in somewhat parallel tracks designed to save time. Some scientists would go to Tinian as experts on the atomic bomb to help set up the technical details and make the military preparations, while others would stay to monitor and carry out the first bomb test. If the test worked, Japan would be attacked with atomic bombs in short order.

The plan was that if problems occurred at Trinity—the code name for the location of the first atomic test in New Mexico—then the same kinds of problems could also materialize in Tinian, where bombs would be assembled from materials shipped there by sea and air just before the missions to Japan were to take place. The scientific team could then address the problems on site without delaying the missions.

One such serious problem occurred at the Trinity site when the bomb was too large, as a result of prior heating, to fit into its shell. The heating occurred naturally because a large (although still subcritical) piece of radioactive material was present in one lump. The radiation created heat, as happens to a much higher degree inside a nuclear reactor. The scientists then cooled the core of the bomb, and it fitted perfectly in its casing. All such problems were worked out quickly, and the bomb was soon ready to be used.[18]

Everything was set at the Trinity site for the first atomic bomb to be tested. But nature did not cooperate. The Manhattan Project employed the best weather forecasters in the country, and they predicted clear weather over Trinity, located in the desert some 50 miles north of the town of Alamogordo ("fat cottonwood"), New Mexico.[19] But the weather was rainy and cloudy throughout the day and night of July 15 and into the early hours of July 16, the date chosen for the atomic test. The scientists had to wait until the clouds cleared, otherwise they could not see the explosion. In addition, a pilot was to fly an airplane in the vicinity of the test at the time of the drop to take pictures and make measurements. At 4 AM the rain stopped and the sky began to clear.

The weather was good enough to carry out the test but presented problems for Captain W. S. Parsons, who was to pilot the B–29 not only to observe and photograph the test from the air but to judge how the airplane responded to the atomic blast. He would then take this information with him to Tinian and Hiroshima. As it turned out, the lingering upper atmospheric clouds defeated this plan—to see the test, he would have had to be 200 miles away, beyond the cloud cover, and thus unable to learn about the effects of the blast on the B–29. Nevertheless, after several brief postponements, the Trinity test proceeded.

The bomb, a plutonium implosion device, the Fat Man (FM) type of weapon, had been hoisted on top of a tower at Ground Zero. Groups of scientists and military personnel observed the test from various locations at different distances from Ground Zero. Oppenheimer and many of the top scientists were at the command location 10 miles away from the bomb, and General Groves and other military observers were at another location about 15 miles away. Norman Ramsey was watching the test with Enrico Fermi on one side and Isidor Rabi on the other: "There were a dozen or more of us, in a line up against a bank, with welders' glasses that had been issued to us—we were also supposed to look in the opposite direction when it went off, through the welders' glasses. I must admit I thought that was going too far," Ramsey recalled, "being quite fearful that I might thereby see nothing whatsoever. As it turned out, it was plenty bright."[20]

On July 16, 1945, at 5:29 AM, the United States exploded the first atomic bomb. Oppenheimer recollected later that the sight of the mushroom cloud over Ground Zero made him think of the Hindu epic *Bhagavad-Gita,* in which Vishnu, a multi-armed deity, says: "Now I am become Death, the destroyer of worlds." Oppenheimer knew that the world would never be the same. As he left the control room of the operation, Kenneth Bainbridge, a physicist from Harvard University, said to him: "Now we're all sons of bitches."[21]

This feeling, however, was not universal. Different scientists who worked on the bomb, under pressure and in wartime, had various reactions to what they were doing and the possible implications of their work. The scientists who watched the first nuclear blast in history, wearing their welders' glasses to protect their eyes, acknowledged that they had witnessed an awesome sight.

What struck many of the observers of the first nuclear explosion was the visual effect, rather than the sound or other effects of the detonation. According to Ramsey, "You could see brilliantly in the opposite direction from it, and turn around relatively slowly before you could look directly even through the welding glasses." He continued,

> My memories are clearest about the visual part of the explosion. In some ways the most astonishing thing to me was that looking in the opposite direction, through these welders' glasses, the hills miles

away were brilliantly lighted. You could look more in the bomb direction only gradually, at the fastest rate you dared come around,
because the brightness of the light was such it wasn't safe to come
faster, but it seemed to me I spent a long time before finally looking at what was by then a fire ball. Again, my recollection is of a
very blue white light.

As Ramsey describes it, the awesome experience surprised them:

Eventually [the fireball] turned redder. By the time we took the
glasses off and looked more directly, it was the very nasty looking,
well-known fire ball which is much nastier to see than it is in a
movie. This fire ball then went up into a mushroom cloud. I don't
know if we were anticipating quite as big a cloud effect. After that
there was a good deal of worry about where the cloud was drifting.[22]

The observers' trousers were ruffled when the shock wave reached
their location some miles from Ground Zero, and the sky glowed purple from the ionization of air molecules that resulted from the intense
radiation. The audible blast and the rushing winds that followed the
shock wave came minutes after they saw the fireball—light, and radiation, travels so much faster than sound and air.

In the tense moments before the test, the scientists talked mostly
about whether or not they thought the test would be successful. No
one had ever attempted to explode such a bomb, so all that they had
to go by were theoretical calculations and extrapolations of the intensity of the neutron flux inside the core based on experiments that had
been conducted at lower intensities.

One person who was outwardly calm but was clearly excited about
the experiment was Enrico Fermi. When the fireball of the explosion
lit up the sky more intensely than the midday sun, Fermi started to
drop little pieces of paper from his hand. He was performing a simple
but highly effective and ingenious experiment—which was quite typical of his work. The air was fairly calm, and the pieces of paper were
dropping straight to the ground. But Fermi knew that, at the location
where he and the other scientists were standing, ten miles away from
Ground Zero, there would be a huge shock wave, and he wanted to estimate its magnitude. He continued to drop his little pieces of paper,

and some moments later, when the powerful wind and sound blast reached them, Fermi estimated the magnitude of this shock wave with surprisingly good accuracy based on how far away the pieces landed.[23] Such elegant experiments gave us the term "Fermi estimates," used for back-of-the-envelope calculations that can in a simple way capture the essence of a physical or mathematical problem and provide good rough answers to what may be complicated questions.

After the blast was over, Oppenheimer's group, who had observed the explosion from the closest location to Ground Zero, gathered with the others, including General Groves. They all worried about the fallout from the bomb, and once they estimated where the cloud was drifting, an army tank, which had been especially fitted with a lead bottom for protection from radiation, was sent in the direction of drift to collect dirt samples that had become radioactive.[24] Weeks later, scientists examining the ground at the blast location itself recovered glass that had formed from the heat of the explosion and that contained significant amounts of radioactive material. Eventually the wide crater with the glass and dust from the atomic blast would be bulldozed over with dirt to contain the radioactive debris.

Oppenheimer, on whom the entire operation and its consequences had depended, was deemed by his colleagues too jittery to drive. So around noontime, after they had all been at the test site since the night before, through the predawn explosion and its aftermath, Norman Ramsey drove Oppenheimer (who had come there in his own car) and Isidor Rabi back to Los Alamos. They were all tense and exhausted, so on the way back did not talk about what they had just witnessed, concentrating instead on the desert scenery.[25]

On July 17, George L. Harrison, an assistant to Secretary of War Henry Stimson for matters related to the Manhattan Project, sent a coded telegram classified Top Secret (now declassified and in the National Security Archive, at George Washington University) to his boss:

War 33556
Secretary of War from Harrison

Doctor has just returned most enthusiastic and confident that the little boy is as husky as his big brother. The light in his eyes discernible from here to high hold and I could have heard his screams from here to my farm.

End

The telegram is a report on the success of the Trinity test. The "little boy" was the untested uranium gun-type bomb; the "big brother" was the Fat Man bomb (plutonium implosion design), which was exploded. The light of the explosion could be seen from "here"—meaning from Washington, D.C.—to "high hold"—which was Stimson's estate on Long Island—that is, a distance of 250 miles away from the test site. The words "my farm" refer to Harrison's farm in Upperville, Virginia, which is 50 miles away from Washington. This meant that the sound of the explosion ("his screams") could carry to a distance of 50 miles.[26] This cryptic message was written in Harrison's encoding using personal details and presumably originated from information conveyed to Harrison in Washington by General Groves.

The following day, July 18, Groves sent his own top secret memo to the secretary of war. While the document was declassified in 2005, parts of it were blacked out, still deemed confidential in the twenty-first century. The lengthy memo is excerpted below:

WAR DEPARTMENT
WASHINGTON
TOP SECRET

MEMORANDUM FOR THE SECRETARY OF WAR.
SUBJECT: The Test.

1.   This is not a concise, formal military report but an attempt to recite what I would have told you if you had been here on my return from New Mexico.
2.   At 0530, 16 July 1945, in a remote section of the Alamogordo Air Base, the first full scale test was made of the implosion type atomic fission bomb. For the first time in history there was a nuclear explosion. And what an explosion! [the next 2 1/2 lines of the memo are censored] The bomb was not dropped from an airplane but was exploded on a platform on top of a 100-foot high steel tower.
3.   The test was successful beyond the most optimistic expectations of anyone. Based on the data which it has been possible to work up to date, I estimate the energy generated to be in excess of the equivalent of 15,000 to 20,000 tons of TNT; and

this is a conservative estimate. Data based on measurements which we have not yet been able to reconcile would make the energy several times the conservative figure. There were tremendous blast effects. For a brief period there was a lightning effect within a radius of 20 miles equal to several suns in midday; a huge ball of fire was formed which lasted for several seconds. This ball mushroomed and rose to a height of over ten thousand feet before it dimmed. The light from the explosion was seen clearly at Albuquerque, Santa Fe, Silver City, El Paso, and other points generally to about 100 [sic] miles away.[27]

The memo continues for 13 pages of technical details and observations about the explosion, the sound level and how far it carried, the light of the explosion being incomparable to anything seen before, the three distinct flashes (as would later be reported in Hiroshima), the nuclear cloud rising into the air, and damage observed at distant sites. Groves quoted in his document a memo from Brigadier General Thomas Farrell in which Farrell discusses the reactions and states of mind of the scientists. Farrell writes that an atomic bomb is no longer just a theoretical construct but a reality, and that America now had "the means to insure its [the war's] speedy conclusion and save thousands of American lives."[28] This last statement has since turned into the battle cry of all who unquestioningly support what happened very shortly after this fateful test in the New Mexico desert.

15

THE DECISION TO USE THE BOMB

The decision of the United States to drop two atomic bombs, on Hiroshima and Nagasaki, at the end of World War II was undoubtedly one of the most controversial acts by any government in history. The scientists working in Los Alamos knew about the Hiroshima attack on August 6, 1945, but some of them were surprised that the second bomb was dropped on Nagasaki only three days later. Eleanor Jette, whose husband, Eric Jette, worked on the Manhattan Project at Los Alamos, recalled: "Our local news of the second bomb-drop was slow coming through. Conflicting reports confused us."[1]

Many of the scientists who took part in the Manhattan Project have expressed various degrees of regret about having aided in unleashing the genie that became the specter of atomic bombs. Einstein, the man whose famous formula made all this—both nuclear weapons and civilian nuclear power generation—possible, distanced himself from nuclear development and the Manhattan Project, even though he was the man who had written to President Roosevelt urging him to launch a nuclear research effort to counter the German atomic development.

But by the time of the Trinity test, the Nazi threat had been over for two months. Einstein is reported to have often said after the war, "Had I known that the Germans would not succeed in producing an atomic bomb, I would not have lifted a finger."[2]

Why wasn't there simply a demonstration, instead of dropping two atomic bombs on Japan within three days? Many believed that if the Japanese had witnessed something like Trinity, they would have surrendered. Others believed that if Japan hadn't surrendered after the fierce battle for Okinawa, which lasted from April to June 1945, with great losses on both sides, it would not surrender without even greater losses.

In June 1945, before the Hiroshima bomb was dropped, James Franck, a Jewish physicist from Germany who had come to the United States to escape Hitler, and the group of scientists he led at the Met Lab in Chicago wrote a memo to the U.S. government urging it not to use the bomb against Japan but rather to demonstrate to the Japanese leadership the immense power of the bomb by exploding it in a desert or on an uninhabited island. The document, called the Franck Report, was delivered on June 11, 1945. On June 12 Arthur Compton, director of the Met Lab sent this report, representing the views of "key members of the staff" of the lab, to Secretary of War Henry Stimson. "The main point of this memorandum," Compton summarized, "is the predominating importance of considering the use of nuclear bombs as a problem of long-range policy rather than for its military advantage in this war. Their use should thus be directed primarily toward bringing about some international control of the means of nuclear warfare. The proposal is to make a technical but not military demonstration, preparing the way for a recommendation by the United States that the military use of atomic explosives be outlawed by firm international agreement." It continues: "I note that two important considerations have not been mentioned: (1) that failure to make a military demonstration of the new bombs may make the war longer and more expensive of human lives, and (2) that without a military demonstration it may be impossible to impress the world with the need for national sacrifices in order to gain lasting security."[3]

LEO SZILARD HAD been one of the first scientists to understand that a chain reaction of uranium fission was possible, and his fear that the Nazis would obtain a bomb had prompted him to become involved in the Manhattan Project.

Yet as early as the spring of 1945, it was clear that the war in Europe would soon end with a Nazi defeat. Szilard recalled that at that time he began to wonder: "What is the purpose of continuing the development of the bomb, and how would the bomb be used if the war with Japan has not ended by the time we have the first bombs?"[4]

Szilard decided to intervene in a political process that he felt might be taking place behind the backs of the scientists and prepared to petition President Roosevelt against the use of an atomic bomb. Albert Einstein wrote Szilard a letter of introduction that Szilard was going to transmit through the president's wife, Eleanor Roosevelt. Mrs. Roosevelt gave him an appointment for May 8, 1945, when he planned to give her the letter for delivery to her husband. But before that meeting could take place, on April 12, 1945, President Roosevelt died and Szilard's memorandum arguing against the use of the bomb remained undelivered.[5]

Once Truman took office, Szilard was faced with further difficulties because not one of his contacts had channels that would lead to the president. He went to the White House nonetheless, and Truman's secretary, Matt Connelly, told him he should make an appointment with James Byrnes, soon to be appointed secretary of state, who was then in South Carolina. Szilard took the train from Washington to Spartanburg, SC. Byrnes was to vet the petition before Truman saw it. Szilard told him why he wrote the memorandum and explained, in addition, that the Russians might be dragged into a nuclear race with America if a bomb was used against Japan.

Byrnes replied that General Leslie Groves, the head of the Manhattan Project, had told him that "there is no uranium in Russia."[6] This was of course sheer nonsense. Byrnes said he believed that if the United States used an atomic bomb against Japan, this would add pressure on the Soviets to remove their troops from Eastern Europe as soon as the war ended. Szilard was incredulous. Byrnes then told him that as someone from Hungary, Szilard would certainly not want Russians to occupy his native land forever. Byrnes held fast to U.S. government conceptions about the Soviets—which would remain unchanged through the almost five decades of the Cold War. Robert Oppenheimer, whom Szilard went to see next in his attempt to prevent the use of the bomb on Japan, brushed him off as well. Oppenheimer believed that use of the bomb was inevitable but that the United States should tell Russian officials (as well as the British, French, and Chinese) before it was used.[7]

When Groves found out that Szilard had given what Groves considered a secret document to Byrnes, he was furious. To calm him down, the scientists working for the Manhattan Project's Met Lab in Chicago decided to convene a committee to study the matters discussed in Szilard's memorandum, and thus the committee headed by James Franck was formed.[8] As noted, that committee recommended that the atom bomb be demonstrated to the Japanese rather than used.

After the Franck Report, Szilard revised his petition. The first version had been signed by 53 scientists working in Chicago; the second draft gained 15 more signatures. Szilard intended to present it to President Harry Truman in July 1945. The petition urged the president not to use the atomic bomb on people before a demonstration was made, in the hope that this would convince Japan to surrender and make military use of the bomb unnecessary. Some contended that the petition should be sent through regular channels and that there should be no attempt to somehow jump rank and give it directly to the president. Szilard agreed and the petition was then sent through to General Groves's assistant, Colonel Nichols, who was to present it to the president. But Truman was then in Potsdam, ready to issue Japan an uncompromising ultimatum to surrender unconditionally and immediately. Perhaps the petition never made it to his hands. When the Hiroshima bomb was dropped, on August 6, Truman had not yet returned to Washington.[9]

The Szilard Petition may not have made it all the way to the president of the United States, but on the other hand, there were several versions of the petition, sent a number of times, and it is unlikely that the president was ignorant of many scientists' desires that the bomb they helped create in order to defend our country not be used against innocent civilians without a warning. In any case, the U.S. government kept the Szilard Petition confidential and only declassified it in 1961. In a television interview with Mike Wallace that year, Leo Szilard said he had been aware of the immense potential of atomic bombs, and this was why he had urged Einstein to write his now-famous letter to Roosevelt suggesting that the United States immediately start its own atomic bomb project. "Hitler would have threatened us to surrender if he had an atomic bomb and we didn't," Szilard recalled to Wallace. Yet, when the chain reaction experiment was successfully completed by Fermi's team in Chicago on December 2, 1941, Szilard had told Fermi: "I think this day will go down as a black day in the history of mankind." He knew, he said, that the United States had absolutely no choice in the matter because of the Nazis and their atomic bomb quest. "We had to build a bomb," Szilard said, "and there was no discussion of the consequences of the work until the German surrender. Then we tried to exercise the responsibility of scientists by trying to convince the government not to use the bomb against Japan. Until the spring of 1945, there was very little conversation [about the consequences of developing a bomb], practically nothing." Szilard maintained that he would have had the same feelings if the United States had dropped the bomb on Germany. "I don't think any of us hated the Germans," he said, "of course many of us disapproved of Nazism, but many Germans were our friends."

Szilard affirmed that in his view there was absolutely no reason for either invading Japan's main island or attacking its cities with atomic bombs. According to his interpretation, the problem Japan had was that America demanded an *unconditional* surrender. There was no reason to believe that surrender—while not "unconditional"—could not have been simply negotiated between the two nations. "We know that the Japanese had been suing for peace," he told Wallace. "Of the people who actually built the bomb, those who worked with me, I believe that 95% agreed with me. What I should have done was to reach Roosevelt about it, but just then Roosevelt died . . ." Szilard lamented the fact that he could not find a suitable intermediary through whom he might have gotten directly to the new president as he and his colleagues thought necessary. Szilard felt that Truman probably guessed what he and others were trying to contact him about, and therefore he referred them to Byrnes, about to be named secretary of state. But the contact with Byrnes led nowhere, and "After all efforts had failed, I drafted a petition and got 60 of my colleagues to sign it. We sent it through channels and we don't know whether it was delivered. Byrnes did not understand what it was about. He thought [the atomic bomb] was just another weapon."[10]

According to what Szilard told Mike Wallace, the Truman administration was interested in using the bomb because so much money had been spent on this project—the Army, he said, "was determined to drop their bomb." He said the arguments put forth against a demonstration were spurious—"It is true that we had only two bombs, but it would not have been very long before we had more bombs to eliminate the risk that they were all duds." Szilard believed that Oppenheimer took it for granted that once the bomb was built, the army would want to use it, and Oppenheimer himself had expressed his reservations about such a possibility. "I know that he thought that we should not drop the bomb without telling the Russians first," Szilard said, "but I think that he knew that if he proposed that we not drop the bomb he wouldn't get far with it." At the end of his television interview, Szilard became even more emphatic about the decision to bomb Hiroshima and Nagasaki. He stated, "I didn't think we accomplished anything [by dropping the bomb]; we accomplished nothing!"[11]

NORRIS EDWIN BRADBURY, a physicist from the University of California at Berkeley who worked on the bomb and would later succeed J. Robert Oppenheimer as director of Los Alamos, recalled having had "tunnel vision" about what he and others did at that time—they had a job to do, and their main concern was to do it well. Bradbury's main concern was that the Trinity test be successful. In an interview in 1976, he said: "The Trinity shot worked. What was the reaction? Everybody and his brother asked you that question, 'What did you think when the bomb went off?', and you're suppose to have great . . . Hell, you just say, 'Thank Heaven the damned thing worked!' or words to that effect."[12]

Clyde Wiegand, a scientist at Los Alamos, when asked in an interview how he felt after the Trinity test, said: "I don't recall any special feelings. I think probably because we knew what to expect. In spite of that it was still, of course, an awe inspiring sight, or an awful sight." When he was then asked how he felt about the decision to bomb Hiroshima and Nagasaki, he responded: "So many times, though, we're asked how could we live with ourselves after participating in such a dastardly development. But I think it would have been only a matter of time. It's a scientific observation that when you put together some uranium or plutonium and you treat it in a certain way . . . it's going to happen. This we can't control or can't have any opinion on." He added, when prodded to comment on the justification for bombing Japan and how the scientists of the Manhattan Project felt about it: "One also has to remember that many of the people at Los Alamos had relatives who were lost in the war, or in prison in the Orient or Europe, and so there's this entirely different feeling than what you might think during peacetime. I think that this weapon should not have been used on the civilian population. I don't see the excuse for that. Maybe we shouldn't get into that [in the interview]. But how can the United States talk about morality around the world, when practically within a few days of having this weapon, the first thing we did was to throw it on the civilians."[13]

When asked if this was a surprise to him, he answered: "No, we knew what was going to happen. Well, I guess we thought we did. How can I say that? We just believed it and accepted it. There was some last minute talk about threatening, and not really using it. As I remember, the people that I was associated with and the way things were going, I don't think we had any illusions about what was going to happen." Wiegand explained that the scientists all knew that arrangements had already been made ahead of time by the military—even before the atom bomb was tested at the Trinity site. The United States was determined to use the atomic bomb on Japan: the airplanes for the mission had been sent to Tinian Island, the components of the bombs had been shipped there secretly by sea and air, and everything was set in motion. Trinity was simply a move to make sure that the device worked. It was not that the bomb was tested, and then a decision would be made whether to use it on a live target or not. The decision to bomb Japan was a done deal, set in motion months in advance, and nothing was going to stop it.

"On the morning after the bomb test I met Oppenheimer at base camp," Wiegand recalled, "he was walking in the opposite direction from the way I was going, but he paused to say, 'Clyde, we have to get some of these over the Japanese cities.'"[14]

When his interviewer asked Wiegand about proposals for a demonstration, an open test to which Japanese generals would be invited to observe, for example, a battleship in the ocean being obliterated by an atomic bomb, Wiegand responded: "Yes, but after all you've been training a bomber crew and they've been dropping dummies out of the same type airplane, maybe the same one, for months, exactly what they were going to do. So I think it was inevitable what they were going to do. Unless Truman would have said no."[15]

But President Truman did not say no, and Hiroshima and Nagasaki were completely destroyed. Clearly some among the scientists who had worked on the bombs were against their use against civilians. But their calls were not heeded.

NORRIS BRADBURY WAS one of the scientists who did not share such sentiments. In a 1976 interview, when asked if the bombings of Hiroshima and Nagasaki changed how he felt about the bomb, he answered: "I have to say probably not. Again, you see, you'll have to look at what I was involved in. It was basically my responsibility, not directly but pretty closely, that these things worked over there [the atomic bombs dropped on Hiroshima and Nagasaki]. If they hadn't, I won't say it was my neck, but we had done all the things we could think of doing to make sure, but we never, of course, had tested these things in flight." His point was that the scientists were all concerned with doing their jobs well and making sure that the bombs they were making did not fail. Perhaps they did not have the time, or may have felt it was not their place, to worry about the morality of killing hundreds of thousands of civilians. And the scientists themselves did not have the information that the government had about Japanese troop movements, Japanese government decisions, and the state of the war in the Pacific.

Bradbury continued: "I don't know how many people sat down and moaned over the dead Japanese, although there were, of course, old rumors of invasion being planned and the Navy had taken the brunt of the Pacific war, and the cruelty of the Japanese, and so on. I guess at that point we weren't thinking very much about being cruel to the Japanese; they'd been cruel to us too. Honolulu, Pearl Harbor. We weren't about to be very sympathetic."[16]

Bradbury was also asked about the Los Alamos scientists who felt, even before Trinity, that using the bomb was not a good idea. Bradbury expressed doubt about the sincerity of such pronouncements: "Oh, yes. I'm not sure how much of that is hindsight, frankly. Frankly, I never heard it debated here." When the interviewer continued to press him by asking whether the group responsible for the contemporary documents that voiced objections to the use of the bomb had contacted him back then, Bradbury responded: "No, not that I can recall. They wouldn't have approached people in uniform anyway, probably. I bet you don't find anybody wearing a uniform on that list, whatever it may be."[17]

John H. Manley, a physicist who worked with Fermi at the Met Lab in Chicago and later in Los Alamos, had this to say when asked whether he had any reflections, at the time, about the implications of dropping atomic bombs on Japan: "I think I did, you know. You never can be sure about the retrospect and so on. But we'd had some discussions here, not very many. And I think that the main thing, my impression of the main hope that most of us had around here [at Los Alamos, where the interview was held], was that this was really going to make war so terrible that something would really have to be done about it, you see. I think that sustained us. And as far as I'm concerned too, it made a lot of the other things that were proposed, such as the Franck proposal and the Chicago group [Szilard's petition] and so on, not very real. I felt that it [the bomb] had to be used in combat and I still feel that it saved a lot of lives, both Japanese and American—that usual argument. But I think that we were aware, and I was certainly aware. There was just no question that this was going to completely revolutionize warfare."[18]

The issue of whether or not there had been serious wide-ranging discussions among the scientists involved with the Manhattan Project about the decision to bomb Japan continues to be one of great interest in the history of science. Norman F. Ramsey, who supervised the assembly of the bombs on Tinian, said: "There was discussion and a rather mixed amount of discussion. It depended a little upon the stage. Actually, in a certain sense, there was rather less discussion at Los Alamos than, in looking back in retrospect, you might have thought would have occurred, and rather less than possibly some people now like to think there was." The location and assignments of the scientists within the Manhattan Project seem to have determined the degree of discussions about the decision to use the bomb. Ramsey thought that several things "reduced the discussion at Los Alamos. Los Alamos was extremely busy right up to the last, for one thing. There was, I think, more discussion of this in some of the places other than Los Alamos and particularly places whose own job got done earlier and then they were sort of waiting and beginning to think about the consequences." The Trinity test made a difference in the Los Alamos discussions, Ramsey said. "After that [Trinity] I left Los Alamos to go to Tinian, and it's been my understanding that in the weeks immediately following the test at Trinity there was considerable discussion of this problem in Los Alamos, after certain of the other tasks had been finished."[19]

Before the bombing of Hiroshima, U.S. conventional bombing of Japan had been very severe. Every week, about 20,000 tons of explosives were dropped on Japan.[20] Estimates of the yield of the atomic bomb—which later proved to be reasonably accurate—had been made at Los Alamos, and the estimates were in the 18,000 to 20,000 tons of TNT range. Some scientists felt that the atomic bomb they were creating was not that different in power from what was already being used in the war with Japan. The atomic bomb would have provided the U.S. military a means of dropping in one shot the amount of explosives that was being used in a week. This could be seen as simply a problem of increasing the efficiency of the normal bombing operations.[21] People may not have understood at the time the immense potency of a nuclear device—which goes far beyond the simple estimate of tonnage of TNT.

According to Ramsey's recollections, the discussions among the Los Alamos scientists centered mostly on trying to find the most effective way to use the bomb to stop the war. He recalled that one of his colleagues had even seriously suggested that the United States wait until it had *ten* atomic bombs and then drop them on Japan all at once, in order to create an even greater fire power. Then, if the Japanese didn't surrender, they would be attacked next with 100 atomic bombs, and so on: "we could go on in numbers of ten or so because that would make an order of magnitude difference for a period of time and might produce a big enough impact to bring an end to the war. I'm sure that at the present time he [the scientist who made this suggestion] would not feel that way, but it shows somewhat the point of view." This might seem a horrific notion, but given the amount of conventional explosives used in the final weeks of the war, Ramsey observed, "In an objective sense, ignoring the subsequent propaganda and the dramatic effect of the bomb, the number of people who would be killed, this was not large compared to the rest of the bombing going on."[22]

Ominously, the suggestion by that unnamed Manhattan Project scientist was, in fact, taken up by the United States after the end of the war. Less than a decade after World War II ended, the United States developed and produced hydrogen bombs that were, indeed, many orders of magnitude larger than the bomb that destroyed Hiroshima.

THE SCIENTISTS AND ordnance engineers on the Manhattan Project produced estimates of all parameters relating to the bomb, many of which turned out to be correct. As noted, they estimated that the bomb's yield was 18,000 to 20,000 tons of TNT; estimates today of the yield of the Hiroshima bomb vary from 10,000 to 20,000 tons with a commonly reported assessment of 15,000 tons. They also estimated the extent of the area that would be damaged at given distances from Ground Zero, the center of the blast. These estimates, too, turned out to be mostly correct.[23] The one inaccurate estimate was how devastating the fire storm produced by the atomic bomb would be. So many fires were started by the explosion that Hiroshima's fire department was overwhelmed and could not respond. As a consequence, buildings damaged by fires that would normally have been extinguished ended up burning to the ground. Hiroshima and Nagasaki were thus more severely damaged than had been anticipated.

The U.S. government had been meticulously planning the nuclear attack on Japanese cities for at least a year before the bombing of Hiroshima and Nagasaki took place. "The decision made in Washington was that all the targets used for the atomic bomb were placed on a reserve about a year or more in advance," recalled Ramsey, "The air force and navy were forbidden to bomb those targets during the war so that they wouldn't be already overbombed." The U.S. wanted to "save" certain Japanese cities specifically for the hellish destruction that could only be caused by an atom bomb. The original list of cities included Hiroshima, Nagasaki, Kyoto, and Kokura. As it turned out, Secretary of War Henry Stimson had a special feeling for the historic city of Kyoto, which included original old temples and a shogun's palace, and he removed Kyoto from the list of potentially doomed cities.

After the bombings, American scientists were involved in assessing the damage and ascertaining the effectiveness of the two bombs. Robert Serber, an American physicist, was sent to Hiroshima and Nagasaki. In a 1967 interview he said: "Bill Penney and I went to Nagasaki and Hiroshima to make a first quick analysis of the damage. Besides the two of us, there was also a medical team from Los Alamos that went and worked at a lot of things in hospitals, and to see what residual radioactivity was left." The two scientists were there primarily to measure as carefully as they could the exact level of damage that was caused to the two cities. Serber explained: "We did things like trying to pick up samples of oil cans and seeing how many miles away they were flattened. We'd collect samples of concrete and we would analyze the strength. We found shadows on the walls, and from the length of the shadow we could sight back and see how high the bomb was when it went off. You had to use your ingenuity to reconstruct what happened and measure the radiation damage and much of the physical effects of the bomb."[24]

Serber stayed in Japan for two months; his reaction to the destruction was: "It was pretty rough. It was quite remarkable how humans organize for self-protection; it's really remarkable how in a very short time you can adjust to almost any situation. Once you get into a situation of complete destruction and damage, in about two days you get used to it and just go about your business and you ignore the human aspect of it pretty much . . . In both cases [Hiroshima and Nagasaki] we were there before the cities were occupied by the Americans." He felt that he was not in personal danger when he was in Japan: " . . . it was quite remarkable. The Japanese seemed to be a very well-disciplined people. The Emperor said they should cooperate with the Americans, and Penney and I wandered all over the city two or three weeks after. Everybody let us . . . nobody ever threatened us. People seemed quite friendly. You couldn't imagine it in reverse in this country."[25] The Americans were, according to Serber, dressed in uniforms rather than civilian clothes; so they could have been viewed by Japanese civilians as members of the enemy's military. Serber had originally been involved in briefing the pilots on how far away they should be when an atom bomb exploded and what the shock wave from the blast would be like. He was supposed to go on the mission to Hiroshima, to take pictures from an accompanying B–29, but didn't because he didn't have a parachute and none could be found for him.[26]

WAS NAGASAKI NECESSARY? It seems that not only was the Truman administration eager to inflict nuclear disaster on the Japanese, it hastened to do it twice. For what was the point of giving the Japanese only three days from the dropping of Little Boy on Hiroshima to the releasing of Fat Man over Nagasaki? The Japanese were not told that they *had* three days to make a decision in a war that had gone on for years. The second bomb was dropped before anyone could ascertain anything about the Japanese readiness to surrender after Hiroshima.

Opposing this viewpoint, we have the expressed belief that a demonstration would have "wiped out the element of surprise."[27] But one wonders why anyone possessing such immensely powerful weapons, while the other side is about to be defeated or surrender any minute, needs an "element of surprise."

Was Nagasaki necessary? The Japanese had just been hit by the most horrible catastrophe ever visited on a city. Was the second bombing needed? Some writers believe that "there is no reason to think that one shock was enough; the Japanese military could and did claim that there was only one bomb, and that it was not so terrible anyway."[28] This of course is a lame argument. The Japanese understood what the bomb did to Hiroshima, and within a few weeks would have comprehended the true enormity of the damage. But they were not given the benefit of time to assess the vastness of the destruction. They were bombed again almost immediately. The fact that the United States waited barely three days before dropping its second atomic bomb on Japan is something that will haunt America forever.

It seems that once the Japanese had rejected the Potsdam Declaration of July 26, 1945, in which the United States, Great Britain, and China demanded their surrender, the consensus thinking in the Truman administration was that Japan would not surrender.

When physicist Norman F. Ramsey first went to Tinian Island in preparation for Hiroshima, he was instructed that "fifty nuclear bombs might be necessary to force the surrender of the Japanese and end World War II."[29]

But the scientists were not the people making these decisions. They had done their work for the government, and it was the government and the military that made the decisions. The president of the United States and his advisors were the ones ultimately responsible for the destruction of the two Japanese cities by the atomic bomb. What do we know about the political process that led to Hiroshima and Nagasaki?

EVIDENCE FROM A SPYING OPERATION

Much of the pivotal material that has recently been declassified about the decision to bomb Hiroshima and Nagasaki at the end of World War II comes in the form of "Ultra Top Secret" documents that were the result of a special spying operation conducted by the Allies on Japanese diplomatic communications. These documents shed new light on a six-decade-old question: Was the atomic bombing of Hiroshima and Nagasaki necessary?

But the first document that provides new information on the workings of the government and how it viewed the development of the atomic bomb is an internal memo by the U.S. secretary of war written in April 1945, four months before the atomic bombs were dropped on Japan. In this now-declassified memorandum he discussed with President Harry Truman on April 25, 1945, excerpted here, Henry Stimson wrote:

1. Within four months we shall in all probability have completed the most terrible weapon ever known in human history, one bomb which could destroy a whole city.
2. Although we have shared its development with the UK, physically the US is at present in the position of controlling the resources with

which to construct and use it and no other nation could reach
this position for some years.

3.   Nevertheless it is practically certain that we could not remain
in this position indefinitely.[1]

This memo demonstrates crucial, sobering facts about what was on
Truman's and Stimson's minds as early as April 1945. The meeting between the two men also included General Groves, who had to use a
back door to the White House to avoid inquisitive members of the
press. This was the first time that Truman, who became president
barely two weeks earlier when Franklin Delano Roosevelt died, was
made fully aware of the Manhattan Project and its soon-to-be-com-
pleted product. While this memo doesn't deal directly with Japan, or
even with the end of the war in the midst of which the meeting between the president and his secretary of war took place, subsequent
notes make it clear that Japan was intended as the target for the first
atomic bombs.[2]

The men were intent on looking beyond the war. One begins to
see here a trend in their thinking—what happens to Japan is of less
consequence in American policymakers' minds than the general state
of world affairs after the war is over. There is emphasis on the need to
try to contain the nuclear potential.

Later that day, April 25, General Groves wrote a memo in which
he stated that he met with the president in order to "disclose to the
President all of the facts with respect to the Manhattan Engineer District." According to the memo, the president asked many questions
about the project, and these were answered by the secretary of war
without much input from Groves. The memo continued: "The President did not keep the report as he felt it was not advisable. A great deal
of emphasis was placed on foreign relations and particularly on the
Russian situation. The President did not show any concern over the
amount of funds being spent but made it very definite that he was in
entire agreement with the necessity for the project."[3]

This thinking has led some experts to conjecture that the ultimate
fate of Hiroshima was at least in part a message to the Soviet Union.

It seems that when Truman became president, General Groves
tried to increase his political influence by using the atomic bomb

project as leverage. As we see from the next document, Groves the
military man was trying to become a statesman and policymaker.
Two days before his meeting with the president, on April 23, Groves
wrote a "Top Secret" 24-page memo to the secretary of war. The
now-declassified briefing paper from Groves to Stimson reads: "The
successful development of the Atomic Fission Bomb will provide the
United States with a weapon of tremendous power which should be
a decisive factor in winning the present war more quickly with a saving in American lives and treasure. If the United States continues to
lead in the development of atomic energy weapons, its future will be
much safer and the chances of preserving world peace greatly increased."[4] This sounds like a political statement, based on unknown
assumptions about the behavior of nations, curiously coming from a
military man with a very limited, specific charge.

Stimson's diary entry from May 14, 1945, further demonstrates
the thinking in the Truman administration at this time, which was
very concerned with Russia rather than the immensely weakened
Japan. The secretary of war described various meetings with State Department officials as well as the British Foreign Secretary Anthony
Eden. He wrote: " . . . my own opinion was that the time now and the
method now to deal with Russia was to keep our mouths shut and let
our actions speak for words. The Russians will understand them better than anything else. It is a case where we have got to regain the lead
and perhaps do it in a pretty rough and realistic way."[5] While the atom
bomb is not mentioned by name in Stimson's (unclassified) personal
diary entry, the allusion is clear. It lends support to the assessment that
politics vis-à-vis the Russians played a major role in America's ultimate
decisions to bomb Hiroshima and Nagasaki.

The obsession with Russia is apparent in records of further meetings within the highest levels of the Truman administration and with
the president himself. In a memo about his meeting with President
Truman on June 6, 1945, Henry Stimson wrote that there was agreement in the Interim Committee meeting the week before:

> a. That there should be no revelation to Russia or anyone else of our
> work in S–1 [the usual codeword for the atomic bomb] until the
> first bomb had been successfully laid on Japan. b. That the greatest

complication was what might happen at the meeting of the Big Three [the planned Potsdam Conference]. He [Truman] told me he had postponed that until the 15th of July on purpose to give us more time. I pointed out that there might still be delay and if there was and the Russians should bring up the subject and ask us to take them in as partners, I thought that our attitude was to do just what the Russians had done to us, namely to make the simple statement that as yet we were not quite ready to do it.[6]

U.S. policy was definitely focused on Russia. Japan by this time was close to capitulating. And in framing its policies, the United States had ample evidence of the mood in the highest levels of the Japanese government from a highly secret source.

AMERICAN AND BRITISH cryptographers had been working hard throughout the Second World War, as well as before hostilities began, to crack the codes the enemies used in their secret communications. Japanese codes were described in colors—the first was the "Red Code" which was broken early on. The British broke the secrets of the Nazi code-making machine named Enigma in a famous project called "Ultra." During the war, the Nazis shared some of Enigma's capabilities with the Japanese, and this led to the more sophisticated Japanese "Purple Code." American specialists with help from the British then went to work on that code. Once it, too, had been cracked, U.S. Army and Navy Intelligence routinely intercepted, deciphered, and translated Japanese military and diplomatic communications, and provided summary analyses of the information to a select few in the highest levels of the U.S. government, including the president.

The "Ultra Top Secret" (sometimes written "Top Secret Ultra") operation of decoding and translating secret Japanese diplomatic messages was code-named "Magic." A number of the Magic communications have been declassified since 2005 (the sixtieth anniversary of the end of World War II). These most confidential of wartime communications shed new light on the chain of command decisions that led to the atomic bombings of Hiroshima and Nagasaki. These documents have only recently been widely accessible.

On July 12, 1945, Magic sent an Ultra Top Secret memo that summarized the analysis of a secret Japanese telegram sent by Foreign Minister Shigenori Togo to his Ambassador to Moscow, Naotake Sato. The telegram reported to Sato the Japanese emperor's decision to ask the Soviets for help in ending the war. Of course, the Japanese had no knowledge at this point that the Soviets had already committed themselves to the Allies to *declare war* on Japan and thus hasten the end of the worldwide conflict.

The Japanese continued with these fruitless moves, making overtures of peace through the Soviets or through other intermediaries instead of directly to the Allies. The Magic memo below, which was delivered to the president, demonstrates that the Truman administration was fully aware in July 1945 that Japan was suing for peace. The 15-page memo (12 pages of which were declassified on June 30, 2005) begins:

ULTRA
TOP SECRET
No. 1204–12 July 1945   WAR DEPARTMENT
                                     Office of A.C. of S., G–2

"MAGIC"—DIPLOMATIC SUMMARY

MILITARY:

1. Japanese peace move: On 11 July Foreign Minister Togo sent the following "extremely urgent" message to Ambassador Sato:
"We are now secretly giving consideration to the termination of the war because of the pressing situation which confronts Japan both at home and abroad. Therefore, when you have your interview with Molotov [the Soviet foreign minister] in accordance with previous instructions you should not confine yourself to the objective of a rapprochement between Russia and Japan but should also sound him out on the extent to which it is possible to make use of Russia in ending the war."[7]

The next day a second Magic communication was sent to the Deputy Chief of Staff following the intercepts of the Togo-Sato telegrams between Tokyo and Moscow. The subject was "Japanese Peace Offer." The analysis contained the chief army intelligence officer's interpretation of possible explanations for the Japanese moves in Moscow:

(1) That the Emperor has personally intervened and brought his will to bear in favor of peace in spite of military opposition;

(2) That conservative groups close to the Throne, including some high ranking Army and Navy men, have triumphed over militaristic elements who favor prolonged desperate resistance;

(3) That the Japanese governing clique is making a well coordinated, united effort to stave off defeat believing (a) that Russian intervention can be bought by the proper price, and (b) that the attractive Japanese peace offer will appeal to war weariness in the United States.

The memo concluded with the chief army intelligence officer's assessment that: "Of these (1) is remote, (2) a possibility, and (3) quite probably the motivating force behind the Japanese moves."[8]

Whichever of the three points was the closest to the truth, clearly the Japanese were trying to end the war. The terms of their surrender they hoped to negotiate in a peace treaty. On May 8, 1945, the Potsdam Agreement had ended the war in Europe through the Nazi surrender. The Japanese were hoping to find a similar way out of the same conflict. To try to save face in defeat, the Japanese rejected the demand for "unconditional surrender." But since they were clearly ready to end the war as early as July 11, 1945, one wonders why the United States did not work to make it easier for the Japanese to come to terms and end the conflict, instead of continuing with plans for an atomic attack on Japan. The usual answer to this question is that the Japanese were "not ready for peace"—which the Magic communications clearly disprove—and that an all-out U.S. invasion of Japan scheduled for November 1945 would have been inevitable, and with it, the "loss of thousands of American lives."

On July 13, another Magic memorandum was delivered. It begins:

MILITARY

1. Follow-up message on Japanese peace move: On 12 July—the day after advising Ambassador Sato of Japan's desire to "make use of Russia in ending the war"—Foreign Minister Togo dispatched the following additional message on the subject, labeled "very urgent": "I have not yet received a wire about your interview with Molotov. Accordingly, although it may smack a little of attacking without sufficient reconnaissance, we think it could be appropriate to go a step further on this occasion and, before the opening of the Three

Power Conference, inform the Russians of the Imperial will concerning the ending of the war. We should, therefore, like you to present this matter to Molotov in the following terms:

"His Majesty the Emperor, mindful of the fact that the present war daily brings greater evil and sacrifice upon the peoples of all belligerent powers, desires from his heart that it may be quickly terminated. But so long as England and the United States insist upon unconditional surrender the Japanese Empire has no alternative but to fight on with all its strength for the honor and the existence of the Motherland. His Majesty is deeply reluctant to have any further blood lost among the people on both sides, and it is his desire for the welfare of humanity to restore peace with all possible speed.'

"The Emperor's will, as expressed above, arises not only from his benevolence toward his own subjects but from his concern for the welfare of humanity in general. It is the Emperor's private intention to send Prince Konoye to Moscow as a Special Envoy with a letter from him containing the statements given above. Please inform Molotov of this and get the Russians' consent to having the party enter the country."[9]

The emperor of Japan was fully aware that the main stumbling block to a peace agreement was his reluctance to leave office and abdicate, and the decrypted communication bears this out. The information, however, is not incompatible with conjecture (1) of the previous Magic memo, that is, that the emperor was interested in peace and was overcoming more resistant elements in his own military. The communication lends further support to the understanding that Japan wanted to end the war and was looking to negotiate the terms of a peace agreement. The Russians did not comply with the Japanese emperor's request and, in fact, would declare war on Japan after Hiroshima in a self-interested move designed to secure some of the spoils of war.

Further intercepted communiqués show that the Japanese government was not enthusiastic about an "unconditional surrender" and the removal of the emperor but still hoped to achieve an agreement to end the war, and wanted to accomplish it quickly. They were ready to talk. In an intercepted July 18 message from Ambassador Sato to Foreign Minister Togo, decoded by Magic, Sato suggested that the Japanese government agree to an *unconditional* surrender as long as the Imperial

House of Japan was preserved.[10] This shows how desperate some Japanese officials were to end hostilities and come to a peace agreement.

The Japanese kept looking hopefully in the Soviet direction for mediation and seemed unwilling to accept the reality that the Russians had no intention of helping them. Perhaps this was why their peace moves did not succeed, but the fact remains that through the Magic operation, the American government was fully aware that the Japanese were indeed looking for a way to end the conflict immediately and were searching for a solution that would allow them minimal face saving—something that is very important in their culture. The Imperial House of Japan still exists today, so its demise was never really an important goal for the United States. Perhaps the United States wanted to give the Japanese an impossible ultimatum—one that Japan could only refuse. The emperor was never an important part of the equation anyway, and there are strong indications that in the end he was actually an advocate for peace.

Then on July 26, Truman, Churchill, and Chiang Kai-Shek, representing China, issued their uncompromising demand that Japan surrender unconditionally—there would be no room for any negotiation at all. The Potsdam Declaration begins with these words: "Following are our terms. We will not deviate from them. There are no alternatives. We shall brook no delay." And it ended with: "The alternative for Japan is prompt and utter destruction."

The Magic memorandum of July 29, 1945, is one of the most interesting decrypted communications in history. This communication of the enemy's internal information reveals how shocked the Japanese government was by the boldness and unyielding nature of the Potsdam Declaration. They had somewhat expected the call for unconditional surrender, but its unexpectedly aggressive tone made the Japanese suspect that the Russians had betrayed them to the Americans and British by revealing to the Allies that Japan was weak and trying to sue for peace in any way it could. What they naturally did not suspect was that the Americans didn't need the Russians to betray their confidential pleas since the United States had broken the Japanese code. The Magic communication began:

TOP SECRET
ULTRA

No. 1221–29 July 1945   WAR DEPARTMENT
Office of A.C. of S., G–2

"MAGIC"—DIPLOMATIC SUMMARY

1. Tokyo "studying" Allied ultimatum: On 28 July Foreign Minister Togo sent the following message to Ambassador Sato:
    "Reference our No. 944.
    "1. What the Russian position is with respect to the Potsdam Joint Declaration made by England, America, and Chungking is a question of extreme importance in determining our future counter-policy. In view of the fact that——[words uncertain, probably meaning: "as it developed, the proclamations issued at previous"] Three Power Conferences—Quebec, Cairo, etc.—were all communicated to Russia [beforehand], I think it would be hard to believe that the Russians were not also aware in advance of the present Joint Declaration.
    "2. Moreover, since as it happened we were awaiting the Russian answer in regard to sending a Special Envoy, the question arises as to whether there is not some connection between this Joint Declaration and our proposal. Obviously we are deeply concerned as to whether there is such a connection, that is to say, whether the Russian Government communicated our proposal to the English and the Americans, and [we are also concerned] as to what attitude the Russians will take toward Japan in the future."[11]
    (All bracketed words and sentences are in the original "Magic" communication.)

The Americans, meeting with the British and the "Chungking" (Chinese) at Potsdam were emboldened by knowing that the atom bomb was ready and could be dropped on Japan any minute, and from previous Magic eavesdropping and intercepts they knew how weak and desperate Japan really was. This is what allowed Truman to push for the uncompromising unconditional surrender of Japan.

A memorandum of a meeting in the Japanese prime minister's office on July 20 shows that some Japanese believed that the Americans thought that Japan's greatest concern was to keep its Imperial House, and that the Americans were not intent on destroying the emperor's position as head of the Japanese nation. These notions were based on a

Japanese naval attaché's contacts with Gero von Gävernitz in early July in Switzerland. Von Gävernitz was an American of German descent who was the personal secretary to John Foster Dulles (at that time not in the U.S. government). The attaché's memo stated: "If Japan is going to continue the war, the country will fall into a terrible situation where the nation is divided, food runs short, etc., and the Japanese population will be reduced by almost half." He ended by saying that he found it "advantageous to keep contact [with the Americans]."[12]

But these contacts were not to last for long and did not bear any fruit despite the clear American knowledge that the Japanese were interested in peace and were looking for a way to achieve it without a complete loss of face and national dignity. Less than three weeks later, Hiroshima was destroyed, while Truman was still on his way back from Potsdam. Three days after that, before the Japanese could even assess the damage to Hiroshima and respond to it by surrendering, Nagasaki was bombed. The planes for this mission left Tinian headed for the Japanese city of Kokura, but bad weather over Kokura obscured their target. They backtracked to their second target, Nagasaki, also mostly obscured by weather. They managed through a cloud break to locate the Mitsubishi steel works and dropped the Fat Man bomb. The bomb killed at least 75,000 people on the spot and many others from delayed radiation effects.

The fact that twice in three days U.S. airplanes could fly to Japan demonstrates how weak the Japanese were at this point in the war. They had no semblance of an air defense system other than sirens and many of their cities had already been pummeled by conventional bombs. This was a nation close to surrender, not one that would have required a full-scale invasion before it capitulated. The argument that the bombings of Hiroshima and Nagasaki "saved thousands of American lives" is questionable.

But nothing was capable of stopping the American juggernaut—not a potential Japanese surrender, not objections from American scientists, and not the strong plausibility that the bloodshed could be ended without the use of such doomsday weapons. Tremendous resources had gone into this unprecedented project of developing and building atomic bombs, and the military had to show something for all the expense and effort. So even though the Nazis, whose push for a bomb had created this race, were now gone, there were two atomic bombs available as a result of the Manhattan Project, and they had to be dropped somewhere. And politicians understood that these bombs could play a role in sending a message to the Russians.

17

THE COLD WAR

As the Second World War drew to a close after Hiroshima and Nagasaki, the Cold War began. Even before the Japanese surrender on August 15 and the subsequent occupation of Japan later that month by U.S. forces headed by General Douglas MacArthur, other events pointed to an international struggle that would dominate the second half of the century.

On August 8, two days after the atomic bombing of Hiroshima, and a day before the second atom bomb was dropped on Nagasaki, the Russians declared war on Japan and attacked Manchuria. Clearly they were looking to the future—the Cold War that was to ensue—and wanted to secure as much territory as possible.

If the United States thought that it could stave off Soviet encroachment by using the atom bomb, it was gravely mistaken. It could plausibly be argued that Hiroshima and Nagasaki actually created the Cold War, and that if the United States had never used the atomic bomb at all, the Cold War would have progressed more slowly or perhaps not at all. As it happened, the Soviets felt a strong need to match American weapons development by launching their own atomic bomb project.

As early as a year before the end of the war and before the building of the first atomic bomb was underway, some scientists understood that an arms race would develop and were looking for ways to prevent it, or at least to mitigate its effects.

On September 30, 1944, more than ten months before Hiroshima, Vannevar Bush and James B. Conant, president of Harvard University, who held advisory positions in the U.S. government's Office of Scientific Research and Development, wrote a memo to the secretary of war, in response to his query about postwar plans for "special projects" (a code word for the Manhattan Project). Their cover letter summarized their main points:

1. By next summer this will become a matter of great military importance.
2. The art will expand rapidly after the war, and the military aspects may become overwhelming.
3. This country has a temporary advantage which may disappear, or even reverse, if there is a secret arms race on this subject.
4. Basic knowledge of the matter is widespread and it would be foolhardy to attempt to maintain our security by preserving secrecy.
5. Controlling supplies of materials cannot be depended upon to control use, especially in forms which the subject may take in the future.
6. There is hope that an arms race on this basis can be prevented, and even that the future peace of the world may be furthered, by complete international scientific and technical interchange on this subject, backed up by an international commission acting under an association of nations and having the authority to inspect.[1]

Almost a year before the atomic bomb was ready, in the midst of the war, these prescient scientists understood the deep implications of atomic weapons proliferation. They were already looking for ways to minimize its poisonous effects on world peace.

In their detailed analysis, Bush and Conant discuss the atomic bomb with far greater specificity:

> Present Military Potentialities: There is every reason to believe that before August 1, 1945, atomic bombs will have been demonstrated and that the type then in production would be the equivalent of 1,000 to 10,000 tons of high explosive in so far as general blast damage is concerned. This means that one B–29 bomber could accomplish with such a bomb the same damage against weak industrial and civilian targets as 100 to 1,000 B–29 bombers.[2]

The authors continue, arguing against the belief that the United States could maintain a monopoly on atomic weapons, and stressing the need for containment and international controls. The lack of such controls eventually brought us the Cold War between the United States and the Soviet Union.

The Soviets soon built their own atom bomb, and the two superpowers embarked on continuous programs of improving and testing nuclear bombs, programs that were expensive, destructive, and disastrous to human health on a global scale. The American tests destroyed Bikini Atoll in the Pacific, and the fallout would often rain down on Oregon and Washington. Testing was later moved to the Nevada desert, from where the fallout could easily reach most locations on the continent. The Russians used the island of Novaya Zemlia in the Arctic Ocean for their nuclear testing program. Later, the French, British, and Chinese joined in. The British tested in Australia; the French utterly destroyed the pristine environment of parts of Polynesia with bomb tests well into the 1970s, and the Chinese did the same at Lop Nor in the Central Asian desert. The amount of radiation spewed into our environment from these senseless explosions is mind-boggling, and the ill effects on people's health will continue to haunt us for centuries.

In the 1950s, Edward Teller, a nuclear physicist who had worked with Enrico Fermi on the Manhattan Project, joined Stanislaw Ulam (1909–1984), a Polish mathematician who had immigrated to the United States, in devising a design for a hydrogen bomb. Its power came from the creation of fusion, a process by which two nuclei fuse together to form a single entity, mimicking what happens deep in the stars. To get this process started requires an enormous amount of energy, which is why it occurs in the blazing centers of stars. Once it gets going, it releases colossal energies that dwarf those of fission bombs. And here again, uranium plays a key role. To ignite a hydrogen bomb, in which hydrogen nuclei fuse together to create helium nuclei, first a uranium or plutonium fission bomb must explode. The Teller-Ulam design answered the technical problem of producing enough heat energy through fission so that the fusion can take place.

In 1952 the United States tested its first hydrogen bomb at Eniwetok Atoll in the Pacific. The Soviet Union followed with its hydrogen bomb three years later, intensifying the arms race. For comparison, a fission bomb produces a destructive power equivalent to that of tens of thousands of tons (kilotons) of TNT (the Hiroshima bomb had the power of about 15 to 20 kilotons, or 15,000 to 20,000 tons, of high-explosive equivalent), while a hydrogen bomb (a fusion bomb) typically is measured in megatons (its power equivalent to that of *millions* of tons of high explosives). The most horribly powerful bombs produced during the Cold War were roughly equivalent to 50 megatons of TNT (for a Soviet nuclear test in 1961). A 15 to 20 *kiloton* bomb destroyed Hiroshima; try to imagine the damage that would be caused by a bomb that is 2,500 times more powerful.

Once again, some of the scientists whose work had led to these developments tried to confront the politicians who exploited their discoveries. In 1955, during the height of the Cold War, many leading scientists signed what became known as the Russell-Einstein Manifesto. This document urged world leaders to abandon the futile and immensely dangerous nuclear arms race and solve problems among nations through diplomacy alone. Later that year, the most important signatory to this letter, and the most celebrated scientist, pacifist, and humanist, Albert Einstein, passed away.

Leo Szilard spent the last years of his life working hard with the Soviet Union to achieve nuclear disarmament. He began in September 1959 by writing a letter to Soviet Premier Nikita S. Khrushchev, urging him to agree to discuss disarmament with the United States. Szilard obtained answers to some of his letters to Khrushchev and was even granted a two-hour meeting with the Soviet leader in October 1960. He also held meetings with Russian diplomats. The thrust of Szilard's efforts was to arrange meetings between American and Russian scientists to discuss possible solutions to political problems, with an emphasis on disarmament.[3] Even earlier, in the late 1940s, through letters to Truman and to Stalin, Szilard tried to arrange meetings to discuss nuclear issues among scientists and nonscientists alike. These earlier attempts did not bear fruit because the Soviets, until the time of Khrushchev, did not respond.

On November 12, 1960, Szilard, along with Edward Teller, participated in a debate, "The Nation's Future," on NBC Television. In the course of the discussion, he said: "America and Russia have one overwhelming interest in common. We both want to avoid a war which neither of us wants. The Russians have been stressing disarmament as a road to peace. We have not been very interested in disarmament up to recently and this will be evident from our discussion, if you listen carefully, we have not done our homework." He continued in this interview: "Disarmament would not automatically guarantee peace. Let's try to visualize a disarmed world. America and Russia, even in such a disarmed world, would be strong enough militarily to dominate their neighbors."[4] He argued forcefully that the number and total power of the bombs that both sides had was way too high, and that there was absolutely no logical reason to continue to build up larger arsenals of such powerful bombs, The Americans and the Soviets had to come to an agreement to stop the madness. His statements had no immediate effect, but Szilard continued to champion the cause of world peace and nuclear-reduction talks between the United States and the Soviet Union until he died in 1964.

Since the end of World War II, the main objective of the U.S. defense strategy has been "deterrence."[5] The basic idea of the policy was to take a strong and unyielding stance against the Soviet Union by emphasizing the development and production of a large number and a great variety of nuclear devices—some land-based, some aboard strategic bombers kept in flight 24 hours a day around the world, and some aboard nuclear submarines patrolling the world's oceans. Under this doctrine, the United States seemed to be continually sending the message to the Soviet Union, "Do not even think of attacking either us or our allies—with nuclear weapons or your large conventional force. If you do, we will attack you with nuclear bombs."

But the Soviets were building up their own very powerful nuclear arsenal, and in fact some of their hydrogen bombs had a higher yield than the largest American bombs. So the deterrence policy of the United States had to now be viewed in a more symmetrical way: "We deter you, but you also deter us." And as the philosophy of deterrence continued to develop, with both sides racing to build more and more bombs, a new and frightening concept appeared that would guide U.S. nuclear defense policy—and that of the Soviet Union—for decades to come: the doctrine of mutually assured destruction (MAD).

According to the MAD doctrine, both the United States and the Soviet Union possessed so many nuclear weapons (in the tens of thousands on each side), and the means to deliver them, that *both sides would be completely annihilated* if an all-out nuclear war were ever to erupt. In keeping with the MAD model, the United States asserted its "readiness at any time before, during, or after a Soviet attack to destroy the Soviet Union as a functioning society."[6]

But what about a first use of a nuclear device? The United States considered the possibility of attacking the Soviet Union first in order to neutralize the enemy's ability to attack the United States. A first use that was powerful enough to knock out all the enemy's missiles, thus preventing their use in a response to a U.S. attack, was seen as a possible step to prevent a devastating attack by the Soviets. This capability required the development of offensive intercontinental ballistic missiles armed with nuclear devices and aimed directly at Soviet silos housing ballistic nuclear missiles.[7] The basic problem of such a nuclear policy is whether and how and under what circumstances the president of the United States can make the very dangerous decision to forcibly disarm (which in this case would mean annihilate) the Soviet Union. Another, alternative strategy of the Cold War was to "weaken the enemy through attrition."[8]

The attrition policy was also pursued by the Soviets against the United States, and encompassed smaller regional conflicts between the two blocks. The conflicts were aimed at reducing the enemy's ability to concentrate on the larger picture of global conflict, to wear down the enemy's resources, reserves, and resolve to continue the global confrontation. The war of attrition between the Soviets and the Americans brought us the Korean War, the Cuban Missile Crisis, the Berlin Wall conflicts, the Vietnam War, and a host of other regional flashpoints that kept the two blocks in constant non-nuclear confrontation. But the danger that any of these regional conflicts would erupt into an all-out nuclear war remained hanging over our planet for four and a half decades.

One of the architects of this policy was Henry Kissinger. In 1957, long before he became President Nixon's national security advisor and later secretary of state, Kissinger, then a Harvard professor, published a best-selling book entitled *Nuclear Weapons and Foreign Policy,* in which he explained this strategy: "We must never lose sight of the fact that our purpose is to affect the will of the enemy, not to destroy him, and that war can be limited only by presenting the enemy with an unavoidable calculus of risk . . . every [military] campaign should be conceived as a series of self-contained phases, each of which implies a political objective, and with a sufficient interval between them to permit the application of political and psychological pressures."[9] When viewed within the context of Kissinger's calculated strategy, the Cuban Missile Crisis of 1961–1962 takes on a whole new and very ominous meaning. The world came hair-raisingly close to a nuclear disaster in the middle of October 1962, when President John F. Kennedy forced the Soviets to back down and remove the nuclear missiles placed on Cuban soil or face a thermonuclear disaster.

Over the following decades, there were some successful attempts to reduce the global temperature by reducing the numbers of nuclear devices available to the two sides. The Strategic Arms Limitation Treaty (SALT) effectively brought about a mutual reduction in the nuclear arsenals of the two sides. But the Cold War continued almost into the last decade of the twentieth century.

Then in 1989, the Soviet Union began to collapse. On June 17, 1992, Presidents George H.W. Bush and Boris Yeltsin agreed to wide-ranging reductions in the nuclear weapons of their nations, from tens of thousands in the possession of each side to a maximum of only 3,500 nuclear bombs by each country.[10] The production of fissile material for use in bombs was virtually stopped as well. Since 1964 the United States has produced no uranium–235 for weapons, primarily because it is harder to separate uranium–235 from raw ore than it is to purify plutonium, which is produced in nuclear reactors in reasonably large amounts. However, even plutonium production has been put on hold since 1988, thanks to reduction agreements by the two sides. (Existing weapons do deteriorate, but at a very slow rate.)

From 1991 to 1992, in a single year, the number of nuclear-tipped missiles carried aboard U.S. submarines dropped by half, from 352 to 176 in the categories of Poseidon (C–3) and Trident I (C–4) strategic nuclear subs.[11] Similar reductions occurred in the numbers of nuclear weapons carried aboard strategic bombers and held in ground-based silos on both sides.

IN 1982 ARTICLES appeared in the professional literature about a new danger of nuclear weapons—as if death and destruction and radiation damage were not enough. Scientists—most prominently Paul J. Crutzen, a Dutch Nobel laureate in chemistry, and later Carl Sagan and others—determined that if many nuclear bombs were exploded anywhere on Earth, it would create what they called "nuclear winter." This would happen when soot and charred particulate matter rose to the atmosphere, shielding solar radiation. As a result, temperatures around the planet would plummet, and crops everywhere would be destroyed. Nuclear winter would cause everything to freeze, there would be nothing to eat, and people and animals would die. It would be the end of our living planet.

This time, science seems to have succeeded in influencing policy-making. When the United States and the Soviet Union realized that there was no way a nuclear war could be won, and that if the bombs started exploding, we would all lose, the thinking of the politicians began to change. The scientific findings made nuclear war unthinkable and pushed the two political adversaries to begin the process of limiting the numbers of warheads around the world. However, it has recently been estimated that if even fifty Hiroshima-size atomic bombs were to go off in a limited war in the Middle East, or between India and Pakistan, that would be enough to cause nuclear winter and the eventual demise of planet Earth. For this reason, even though the confrontation between the West and the Russians is mostly over, we are still in danger from regional nuclear nations.

URANIUM'S FUTURE

U ranium poses a unique set of risks and benefits as nations both large and small face tough decisions about proliferation, the use of nuclear energy, and the use of uranium's by-products in medical research and treatment.

The twenty-first century must be the age in which we make nuclear proliferation ancient history. Tools exist today for detecting even minute levels of radiation in the environment, and governments must come together to put a complete halt to the illegal trade in nuclear materials. Research must be undertaken to develop new ways to detect nuclear activity by rogue nations.

Nations also need to address vigorously the issues of nuclear waste. We must make sure that the energy derived from uranium in civilian nuclear power-generating plants is clean and safe, while we continue to work on the development of long-range energy alternatives and to design new kinds of nuclear power plants that will offer close-to-absolute safety from accident or terrorism.

THE ATOMIC AGE came about because of the wartime drive to make an atomic bomb. But a side benefit (if one chooses to view it as such) was the scientific understanding of how to use nuclear material to generate electric power. Back in 1942 Enrico Fermi's experiment at the University of Chicago demonstrated that a chain reaction of uranium—not necessarily the pure 235 isotope—could be sustained. Generating power does not require many of the complicated steps necessary to make a bomb. Here is how the chain reaction process works in power generation.

Uranium is refined somewhat, so that it contains more than the natural percentage of the 235 isotope but at a far lower level of enrichment than that necessary for bombs. It is sufficient to have uranium that is mostly the 238 isotope with a small percentage of the rarer uranium 235. This slightly enriched mixture of uranium isotopes is used as the fuel in a nuclear power-generating plant. The main part of the plant is the reactor: a large insulated vessel that is well shielded for radiation leakage and sits inside a containment dome, which adds a level of protection.

The reactor contains fuel rods made of pellets of enriched uranium. Uranium atoms undergo fission, releasing neutrons, which in turn impact other uranium atoms, which then undergo fission. This is the same chain reaction achieved by Fermi and his group in the famous 1942 experiment. The uranium chain reaction inside the reactor is moderated using heavy water, regular water, graphite, or another material that absorbs excess neutrons to prevent the reactor from going supercritical and exploding. Cadmium rods are inserted into the reactor to keep the reactions at controlled levels.

In the nuclear process inside the reactor, heat is one of the products, as explained by Einstein's equation relating mass to energy. The mass of uranium changes as it undergoes fission in a chain reaction, and heat is part of the energy that is released. The purpose of the nuclear power plant is to exploit this heat energy to run a steam turbine. The heat from the nuclear reactor is used to boil water, or to heat it to a high temperature in some designs. The heat energy is then turned into kinetic (motion) energy. In the case of a steam-generating plant, the steam produced from the boiling water goes inside a turbine and makes it turn. This is the same principle as in a steam locomotive or a steamship, except that instead of coal or other fuel, it is nuclear power—the continuous chain reaction of uranium fission—that produces the heat energy to run the turbine.

The kinetic energy, the motion of the turbine, is in turn converted to electric power. This is done by using the kinetic energy derived from the steam to spin large coils of copper wire in a magnetic field. When metal wires are made to move inside a magnetic field (created by large magnets) in a generator, the result is electricity—the forced flow of electrons in the wire. Thus uranium fission leads to the production of electricity. This process is very efficient: The total heat generated by a slowly decaying quantity of uranium is greater than might be produced by the burning of fossil fuels of far greater amounts. But nuclear power brings with it special problems.

Once the pellets of enriched uranium are spent, they contain a variety of radioactive elements that are no longer useful in generating heat. They must then be removed from the reactor and replaced by new fuel. This is when the reactor shuts down for maintenance and refueling. But the spent fuel is highly radioactive and many of the products of the nuclear reactions remain radioactive for many thousands of years. So a permanent solution to the safe disposal of spent nuclear fuel must be found. To date, only temporary measures have been used: storage on site and shipment to nuclear waste facilities both offer disposal that is not permanent.

The Russians and the British were the first nations to take advantage of the new technology for power generation. In 1954 the Russians connected a small research reactor at Obninsk, southwest of Moscow, to the public power grid, thus adding a small amount of nuclear-produced electricity, about five megawatts, to that obtained through conventional means.

The British government had been using a facility called Sellafield, in Cumbria, in the northwest of England, for nuclear research to produce weapons. On this site in 1956 the government inaugurated the Calder Hall reactor, which converted the nuclear processes used in research to large-scale production of electricity for public use. This reactor provided ten times the flow of electricity produced by the first Russian reactor. The reactor was closed down in 2003, after almost five decades of operation.

Rather than simply store the spent uranium from Sellafield, the British tried recycling. Since plutonium is produced in nuclear reactors, and plutonium can also be used as a fuel for reactors, they saw an opportunity to save resources. The idea is to reprocess the spent uranium fuel of nuclear power plants, extracting from it the reusable plutonium, which can then be used as new fuel for electricity production. The British built in their Sellafield facility a nuclear reprocessing plant that carried out this task. But reprocessing spent nuclear fuel is a difficult and dangerous task. The spent fuel contains other highly radioactive elements in addition to the usable plutonium, and these must be extracted and removed in a clean way that will not lead to contamination. As it turned out, the Sellafield reprocessing facility has been plagued by problems of release of radiation into the environment. So far, reprocessing generally has not delivered on its promise.

In December 1957 the United States opened its first nuclear power-generating plant, in Shippingport, Pennsylvania, and after a test period of running on low power, it went on line at 60 megawatts of electricity in 1958. The reactor worked for a quarter of a century; it was decommissioned in 1982 and subsequently dismantled. Shippingport was the first reactor built under President Dwight Eisenhower's initiative, announced in 1953, to inaugurate broad-scale nuclear power production in the United States. The nation has since built over 100 nuclear reactors, roughly 2 per state, 104 of which are still in operation today. Nuclear energy has been viewed as a clean and relatively inexpensive power source, and as a result, the United States now generates close to 20% of its electricity from nuclear energy.

At the end of the war, in 1945, Frédéric Joliot was appointed director of France's new Atomic Energy Commission, and under his guidance France developed both military and commercial atomic energy. The French were eager to forge ahead with nuclear development and began testing atomic bombs in the Sahara, in their Algerian colony. The first French nuclear bomb tested had a yield of 70 kilotons—three or four times the yield of the Trinity or Hiroshima bombs. After building a powerful nuclear-armed military, the French turned to the construction of nuclear power-generating plants. This move toward atomic energy intensified in the 1970s, and at present

France maintains 59 working reactors, providing as much as 80% of the electric power generated in that country. (The exact percentage is somewhat hazy and became an issue in the last French presidential campaign, when neither candidate knew the actual percentage; all agree it is not much less than 80% and is the highest percentage of any nation in the world.) With such a high level of total electricity produced by nuclear power plants, France is the world leader in exploiting nuclear energy for peaceful purposes, and it even exports nuclear-generated power to other nations.

Nuclear electric power was also developed by other countries, including Japan, China, and many European nations. Overall, there are 439 nuclear power reactors operating in the world at present, providing about 15% of the total electricity produced on the planet. The largest producers of nuclear-generated electricity are the United States, France, and Japan, which at present has 55 nuclear power-generating reactors in operation. Large reactors produce hundreds of megawatts of electricity, and the most powerful ones have a yield of close to 1,500 megawatts. For decades, it seemed that the world was moving closer to a high dependence on nuclear energy as a main source of power.

In addition to nuclear plants that generate electricity for public consumption, as well as the many small nuclear research reactors (typically yielding a few megawatts of power) operated by the military and educational and research institutions, there are submarines, aircraft carriers, and even smaller vessels such as icebreakers that use nuclear technology for propulsion. The reactors aboard ships and submarines are necessarily much smaller than the land-based power-generating reactors; they make up for their small size by using more powerful fuel. Typically, nuclear fuel for naval vessels contains a much higher proportion of uranium 235 than do electrical power stations, although the level is still much lower than in weapons-based uranium. These reactors are housed where a ship's engine would normally be located, and they work by producing high-pressure heated water and steam that generates motion along with electricity. These are, in effect, miniature nuclear power plants. The advantage of using nuclear propulsion in a submarine is that the vessel can stay submerged for long periods of time and requires refueling only after several years of operation. With the end of the Cold War, the need for such nuclear-powered ships is decreasing. In fact, the Russians now use some of their docked nuclear-powered naval ships to supply electricity to remote towns in the arctic.

What is the future of nuclear power generation? Over the decades and around the world, many accidents have occurred and continue to occur in nuclear power plants and research reactors. These range from shutdowns of power stations due to small nonnuclear-related problems to major accidents that result in contamination of nuclear workers or even the general population. Both the British and the Russians had unpublicized problems with their nuclear technology over the first few decades of its use. Then in 1979 the first major media-publicized accident occurred.

At 4 AM on March 28, 1979, a reactor designated Unit 2 of the Three Mile Island Power Station near Harrisburg, Pennsylvania, suffered a small technical problem, which later deteriorated to a leakage of the water cooling the reactor. The management of the plant failed to realize what was happening and did not take corrective action. As the water continued to leak, the reactor temperature rose precipitously. Eventually, the heat of the reaction vessel reached a level at which a partial meltdown occurred. Because part of the reactor was damaged by the melting of the metal containing the radioactive material in the reactor's core, radiation leaked into the environment.

American reactors have a built-in level of added protection: a containment building that covers the actual reaction vessel. So when the radiation began to leak out, much of it remained contained in the building. In the final analysis, once the situation came under control and the leakage was stopped, the total amount of nuclear contamination was relatively small. Cancer rates in this area are not considered statistically higher than elsewhere. The Three Mile Island accident, nevertheless, was a milestone in the history of nuclear power generation because it immediately raised public concern about safety. Seven years later a reactor failure on the other side of the world would be much worse.

In the early morning hours of April 26, 1986, nuclear workers at a power plant near Chernobyl, Ukraine, shut down the emergency warning system of reactor 4 to conduct a series of tests. Without the warning system, they could not tell what was happening, and something they did in their testing caused a giant explosion. The entire reactor burst into flames, spewing its contents into the air. Since the Russian-designed reactor did not have a secondary containment building on top of it, the highly radioactive contents of the reactor went into the atmosphere. To add to this disaster, the Soviet authorities decided to keep the accident secret. They even allowed a large public parade celebrating the May Day holiday to take place in the area, thus exposing many thousands of unsuspecting citizens to high levels of radioactivity.

The incident came to light days later when engineers at a nuclear power station in Sweden—800 miles to the northwest—noticed higher-than-usual radiation measurements in their instruments. After determining that the radiation did not emanate from their own plant, they deduced something strange was happening in the atmosphere. Further tests indicated that a cloud of radioactive contamination had spread over the region from the southeast. Only after further investigations by authorities in the West, all indicating radioactive contamination originating from the Soviet bloc, did the Ukrainian government acknowledge the disaster. By then, much of Eastern and Western Europe had been affected by radiation, as well as parts of Asia and North America.

The Chernobyl accident released planet-wide radioactive contamination believed to have been hundreds of times as severe as the fallout from an atomic bomb. It was the worst publicly known nuclear accident in history—although a mysterious explosion in the Ural Mountains in 1957, kept secret by the Russians, may have come close to the Chernobyl damage. The Chernobyl disaster is believed to have resulted in thousands of cancer cases, mostly in the immediate region. Since the plant is located near the border with Belarus and prevailing winds were blowing in that direction, much of Belarus was affected and cancer rates in that country rose significantly following the explosion in Chernobyl.

In the wake of Chernobyl, Italy decommissioned all its nuclear reactors, and many countries halted—at least temporarily—the construction of new nuclear power plants while waiting for more comprehensive decisions to be made, based on new scientific evidence about the safety of this powerful source of energy.

It's been more than two decades since the accident, and some countries are now returning with new vigor to embrace nuclear energy. The Swedish government announced in February 2009 that it planned to replace ten of the country's aging nuclear power plants with new and more efficient models. And there is growing interest in nuclear power in many countries; some are reevaluating their stance on atomic energy. There are many reasons for the shift back to pre-Chernobyl days.

The resurgence of interest is driven in part by the problem of global warming. Burning fossil fuels such as gas, oil, and coal produces large carbon emissions. When the carbon dioxide that is produced as a result of burning fossil fuels in power plants rises into the atmosphere, it contributes to the greenhouse effect that keeps thermal radiation from escaping back into space. Thus, our planet heats up and temperatures generally rise. Nuclear reactors can help reduce the extent of this problem. Since their power source is uranium fission, which does not create carbon dioxide, such plants do not emit greenhouse gases. Nuclear power, along with alternative power sources such as wind, solar, and hydroelectric energy generation, can help combat global warming.

Other pollution problems result from the operation of fossil-fuel power plants. In addition to carbon dioxide, these power plants emit gases such as sulfur oxides, which are known carcinogens, as well as toxic metals such as mercury. Mercury results from coal burning and contaminates fish in lakes, streams, and oceans, causing significant health problems, and posing special dangers to pregnant women as it can harm fetus development. Oxides from coal burning produce acid rain, which kills entire fish populations in lakes. Safe nuclear power could help avoid these problems.

In addition to global warming and pollution problems that result from burning fossil fuels, we face another problem: the unstable price of oil and its limited global supply. Much of the world's oil is produced by countries whose interests do not always agree with those of consuming nations. The dependence of the West on the oil resources of nations in the Middle East, Africa, and South America produces economic instability as well as the possibility of political pressures, as exemplified by the oil embargo of 1973. The further development of nuclear technology can help mitigate these problems.

Nuclear reactors are expensive to build in part because of stringent safety controls required in most Western nations. But once the initial capital investment in a plant is made, running it is generally less costly than running a plant that uses fossil fuels, especially over the thirty-year lifespan of an average nuclear reactor. Additionally, nuclear fuel can be recycled to make plutonium, which can then be reclaimed from the nuclear waste. But the recycling requires special plants, and these are considered more dangerous than nuclear power plants. And they are expensive to run. Additionally, there is always the danger posed by nuclear waste.

The lure of the atom is great—there is much that can be achieved by devising new ways to ensure the safety of nuclear power plants. Even without the currently costly and dangerous recycling of nuclear fuels, the world still possesses limited but large deposits of uranium ore, which can be exploited in a safe system of nuclear power plants. Safety is the issue of paramount importance here. If power plants can be made perfectly safe, then the use of the uranium resources can help bring us a future of clean energy and reduced dependence on foreign oil.

In addition to designing a new generation of safer power plants, we must also find a permanent way to store nuclear waste. Just as radiation causes damage to people and other living creatures, it damages materials, too. A steel drum containing radioactive waste will slowly corrode over decades, both because of erosion in the ground and because of radiation damage from within. Eventually, the drum will leak, and the radioactive isotopes will find their way into ground water or streams or the ocean, causing radioactive contamination. When thinking about radioactive waste disposal, we must plan ahead with a far longer horizon than decades or even centuries. We must think thousands of years ahead, something policymakers are not used to doing.

The future will tell whether scientists can find ways to solve the serious problems posed by nuclear waste—which, with isotopes that have half-lives in the many thousands of years, remains radioactive practically "forever." If these problems can be solved and safe, clean energy can be produced, we may be able to use nuclear power stations to help solve the present global warming issues as well as provide affordable electric power to billions of people.

THE WORK OF Marie and Pierre Curie—especially of Pierre Curie, who worked directly with doctors—initiated the use of radiation in cancer treatment. The earliest treatments involved injecting tumors with radium-filled needles. Radium is so much more radioactive than uranium, it intensely bombards the surrounding tissue. The powerful radiation shrinks tumors and even dissolves them completely. Eventually the cells near the implant become so damaged by radiation, they die and the cancer disappears. The problem is, of course, that it is hard to kill all cancer cells and simultaneously minimize the death of normal cells.

In addition, radiation can cause cancer in the same way that it can kill cancerous cells. If a DNA helix in the cell's nucleus is hit by a ray or particle, an electron is stripped off an outer orbit in an atom, and this can break the DNA chain, causing loss of information. Later, when the cell divides, it does so in a defective way because its DNA is damaged. This can lead to both cancer and genetic defects (in the case of a reproductive cell). So radiation therapy, as is true for all ionizing radiation, can also cause cancer and birth defects. As patients treated with radiation, however, already have cancer, the good usually outweighs the bad.

In the 1950s, once civilian and military nuclear reactors were producing many kinds of radioisotopes in their cores, scientists found excellent replacements for the expensive radium, whose cost had been as high as tens of thousands of dollars per gram. A gram of cobalt 60, an isotope produced in reactors, provides an amount of radiation that is equivalent to that of a gram of radium but costs only about one dollar to produce. Cobalt 60 and other radioactive elements produced in reactors have now completely replaced radium for use in medicine.

Radioisotopes are also used in diagnostic medicine. Since these elements "glow" (their radiation is detectable and can be traced by sensing devices), they are used in many kinds of medical diagnoses. One example is the use of technetium (a radioactive element produced in laboratories) to aid doctors in observing the flow of blood in the veins of people suffering from blocked arteries. Stress tests of heart performance often employ radioactive tracers that shortly afterward leave the body in human waste.

THE END OF the Cold War did not bring about clear policies on how the sides should deal with each other, especially on issues of nuclear weapons and nuclear proliferation. In a book that surveys many recent publications on U.S. global post–Cold War policy, Hans Binnendijk and Richard L. Kugler conclude that the end of the Cold War brought about four distinct U.S. political viewpoints. Traditional Conservatives, according to this survey, believe in classical power-balancing military relationships, that is, a continuation of the use of military strength as a deterrent as well as a preserver of the status quo achieved with the Russians. Progressive Multilateralists believe that the United States should act throughout the world, in concert with other nations, as the leader of the democratic nations on our planet. Assertive Interventionists hold that the United States must assert its military and economic power unilaterally across the globe and deal with global threats without necessarily consulting other nations. And finally, followers of the Offshore Balancing philosophy believe in withdrawing U.S. troops from most offshore locations in the world and follow a neo-isolationist view; this idea is held by some on both ends of the political spectrum.[1]

But while great progress was made during the 1990s and into the twenty-first century in reducing the potential for a nuclear war and improving the relations between the United States and Russia following the collapse of the Soviet empire, the late years of the administration of George W. Bush were marked by a dangerous regression to Cold War conditions. Former Soviet satellite nations, which had been members of the Soviet Union's defense treaty, were courted for membership in the American-led NATO. This move angered the Russians, who saw a new potential military threat at their door. In addition, in 2008 the Bush administration decided to place strategic nuclear missiles in Poland, ostensibly as a defensive threat against Iran, thus further frustrating the Russians.

Cold War suspicions were evident also in the U.S. Supreme Court's November 12, 2008, ruling that overturned decisions by California's lower courts and lifted restrictions that had been imposed on the U.S. Navy's use of sonar in exercises in the Pacific in the vicinity of pods of whales. The whales' sensory organs, nervous systems, and brain functions are severely harmed by the use of sonar aboard submarines. The whales often beach themselves and die because of the damage to their bodies caused by the sonar. Yet the Supreme Court ruled that the Navy's need to conduct such exercises overrides the well-being of these endangered species. In an age that has seen deep cuts in the numbers of nuclear submarines on both sides—the old Soviet submarine fleet has been rusting on Arctic shores for over a decade and the American fleet has been greatly reduced in size—one wonders why and how there is a perceived need for new sonar exercises aimed at detecting a hostile submarine attack.[2]

The end of the Cold War signaled the arrival of an era of proliferation of nuclear weapons beyond the largest nations. Pakistan began its nuclear program in the 1970s, and India and Israel also belong to this third generation of nuclear nations. The leading Pakistani nuclear scientist, Abdul Qadeer Khan, has sold nuclear secrets to a number of other nations: Libya, Iran, North Korea, and perhaps others.

Khan, an expert in metallurgy, had reproduced the Manhattan Project in Pakistan and made improvements in the process of extracting uranium 235 from ores containing a predominance of the heavier 238 isotope. The Iranians obtained his knowledge of how to use fast-spinning centrifuges to effect this separation. Today the future of Pakistan's nuclear weapons program is uncertain, as is that of North Korea. That many nuclear warheads of the former Soviet Union are unaccounted for makes for more global unease.

World security is imperiled by the proliferation of nuclear weapons and nuclear materials that can be used in bombs of various kinds (including a "dirty bomb"—one that does not produce a chain reaction but spreads radioactive elements over a large area such as a city). At present, the greatest danger comes from Iran, which has demonstrated that it has the delivery means for atomic bombs. Its ballistic missiles are known to have a range that reaches well into Western Europe, including major cities such as Paris. The fact that Iran is involved in centrifuge work to separate uranium 235 from ore proves that it is not interested in peaceful nuclear power but in a bomb. As explained above, fuel for reactors is not pure uranium 235, but rather a mixture that has a good proportion of the common U–238 isotope. The Iranians could easily obtain such material on the open market. Their intensive refinement operation, which produces pure U–235, only used in atomic bombs, would be immensely expensive overkill to use in a reactor that would work with other fuel.

It cannot be stressed enough that the MAD doctrine, which saved us from nuclear holocaust on many occasions during the Cold War— during the Cuban Missile Crisis, in Vietnam, in Korea, and at other flashpoints throughout the forty-five years of the East-West conflict— will not necessarily work in the case of Iran. And it will certainly not apply in the disastrous eventuality of terrorists obtaining a nuclear weapon from one of the rogue states on this planet. Suicidal behavior brought on by fanaticism cannot be deterred by that old logic.

If we can solve the difficult worldwide problems of nuclear proliferation, nuclear waste, and nuclear safety, uranium could play a positive role in our future. Within a system of nuclear reactors constructed to operate without risk of leakage of radioactive waste, the power of uranium could be harnessed to produce clean energy. Radioisotopes, nuclear fission's by-products, may continue to be useful in medicine, helping in diagnoses of disease and providing cures for cancer, both directly and through medical research. The end of the Cold War and the changes in the global economy place us at a crossroads. Now is the time to make the important decisions about uranium and its uses— decisions that can make a difference in the future of our planet.

NOTES

INTRODUCTION

1. Mikio Kanda, ed., *Widows of Hiroshima: The Life Stories of Nineteen Peasant Wives,* trans. Taeko Midorikawa (New York: St. Martin's Press, 1989), p. 5.
2. Mikio Kanda's interview with Setsuko Nishimoto, in Kanda, *Widows of Hiroshima,* p. 5.
3. Mikio Kanda's interview with Setsuko Nishimoto, in Kanda, *Widows of Hiroshima,* p. 6.
4. Mikio Kanda's interview with Tsuruyo Monzen, in Kanda, *Widows of Hiroshima,* p. 21.
5. Kanda, *Widows of Hiroshima,* p. xiii.
6. Y. Tsutomu et al., *Journal of Radiation Research* 28 (1987), 156–71.
7. Kenzaburo Oe, *Hiroshima Notes* (New York: Marion Boyars, 1995), p. 23.

CHAPTER 1

1. Otto Frisch, *What Little I Remember* (New York: Cambridge University Press, 1979), p. 57.

CHAPTER 2

1. James S. Trefil, *From Atoms to Quarks: An Introduction to the Strange World of Particle Physics* (New York: Scribner, 1979), p. 12.
2. Trefil, *From Atoms to Quarks,* p. 12.
3. Trefil, *From Atoms to Quarks,* p. 13.
4. Trefil, *From Atoms to Quarks,* p. 14.
5. To find this number, we solve for N, the number of months, the equation: $A(0.99)^N =0.5A$, where A is the initial amount (in this case, \$100). Of course the A cancels on both sides (it is placed here just by way of explanation), and we are left with: $(0.99)^N =0.5$, which is solved by taking logs of both sides (it doesn't matter to what base, as long as the base is the same for the entire equation). This gives: $N=\log(0.5)/\log(0.99)$, which, to the nearest integer, is 69. This calculation is similar to many used in financial analyses of mortgages or annuities or other interest-related problems in finance and economics.
6. Isaac Asimov, *Inside the Atom* (New York: Abelard-Schuman, 1966, 3rd ed.), pp. 97–8.
7. Asimov, *Inside the Atom,* pp. 99–100.

CHAPTER 3

1. Ruth Lewin Sime, *Lise Meitner: A Life in Physics* (Berkeley: University of California Press, 1996), p. 10.
2. Sime, *Lise Meitner*, p. 14.
3. Sime, *Lise Meitner* p. 18.
4. Sime, *Lise Meitner*, p. 19.
5. Sime, *Lise Meitner*, pp. 22–3.
6. Sime, *Lise Meitner*, pp. 26–7.

CHAPTER 4

1. Ruth Lewin Sime, *Lise Meitner: A Life in Physics* (Berkeley: University of California Press, 1996), p. 47.
2. Sime, *Lise Meitner*, p. 47.
3. Sime, *Lise Meitner*, p. 48.
4. From our modern perspective, these transitions are easily understood. When thinking about such radioactive decay processes, ignore completely the shell of electrons in the atom around the nucleus—these are *nuclear* processes, which occur inside the nucleus only; the electrons in the outer part of the atom take care of themselves, that is, they are added or removed through the environment. And we completely ignore the number of neutrons in the nucleus, as they do not affect chemical properties. In any case, we assume that the chemical properties of the element produced are determined solely by the number of protons in the nucleus. Thus, for example, an element with one proton in its nucleus (we ignore the outside electrons) is hydrogen; an element with two protons in the nucleus is helium; three protons gives us lithium, and so on. In the first nuclear process described above, alpha decay, an element loses two protons in the disintegration, hence it becomes (chemically) an element two places down in the periodic table—because its atomic number is now two less than it was before. In the second nuclear process above, beta decay, a *nucleus* loses an electron. This happens when a neutron inside the nucleus disintegrates, losing an electron and thus turning into a proton. The element thus gains a proton, its atomic number (determined by the number of protons only, not neutrons) goes up by one, and it becomes an element located one place *up* in the periodic table. Of course, this was not understood quite this way at that time, since the neutron was discovered nearly twenty years later, in 1932, by James Chadwick.
5. Sime, *Lise Meitner*, p. 58.
6. Sime, *Lise Meitner*, p. 58.
7. Sime, *Lise Meitner*, p. 63.
8. Sime, *Lise Meitner*, p. 64.
9. Sime, *Lise Meitner*, p. 71.

CHAPTER 5

1. Emilio Segrè, *Enrico Fermi, Physicist* (Chicago: University of Chicago Press, 1970), p. 1.
2. Laura Fermi, *Atoms in the Family: My Life with Enrico Fermi* (Chicago: University of Chicago Press, 1954), p. 19.
3. Letter from Adolfo Amidei to Emilio Segrè, reprinted in Segrè, *Enrico Fermi, Physicist*, p. 8.

4. Letter from Adolfo Amidei to Emilio Segrè, reprinted in Segrè, *Enrico Fermi, Physicist,* p. 9.
5. Segrè, *Enrico Fermi, Physicist,* p. 15.
6. Segrè, *Enrico Fermi, Physicist,* p. 18.
7. Segrè, *Enrico Fermi, Physicist,* p. 16.
8. Fermi, *Atoms in the Family,* pp. 37–8.
9. Segrè, *Enrico Fermi, Physicist,* p. 37.
10. Segrè, *Enrico Fermi, Physicist,* pp. 40–1.
11. Segrè, *Enrico Fermi, Physicist,* p. 52.
12. Fermi, *Atoms in the Family,* pp. 60–1.
13. Fermi, *Atoms in the Family,* pp. 44–5.
14. George Gamow, *Thirty Years that Shook Physics: The Story of Quantum Theory* (New York: Doubleday, 1966), pp. 140–1.

CHAPTER 6

1. Emilio Segrè, *Enrico Fermi, Physicist* (Chicago: University of Chicago Press, 1970), p. 71.
2. Segrè, *Enrico Fermi, Physicist,* p. 72.
3. Alpha particles are quantum entities (the elements of the world of the very small: electrons, protons, atoms, and the like), and therefore exhibit properties of *both* particles and waves. Thus we can interchangeably call them alpha *rays* or alpha *particles.*
4. Forty years later, in a physics laboratory at the University of California at Berkeley, my fellow physics students and I repeated this historical experiment. We lowered aluminum cylinders attached to long strings into a small "reactor" that contained an alpha-radiation source, and after pulling out the aluminum element placed it next to the detection tube of a Geiger counter. We were thus able to observe how the induced positron radiation decayed exponentially in a matter of several minutes.
5. Emilio Segrè, *From X-Rays to Quarks: Modern Physicists and Their Discoveries* (Berkeley: University of California Press, 1980), p. 198.
6. Laura Fermi describes her husband's reasoning very beautifully in her memoir, *Atoms in the Family: My Life with Enrico Fermi* (Chicago: University of Chicago Press, 1954), p. 83.
7. Otto Frisch, *What Little I Remember* (New York: Cambridge University Press, 1979), p. 87.
8. Fermi, *Atoms in the Family,* p. 85.
9. Fermi, *Atoms in the Family,* p. 89–90.

CHAPTER 7

1. Laura Fermi, *Atoms in the Family: My Life with Enrico Fermi* (Chicago: University of Chicago Press, 1954), p. 118.
2. Lucy Dawidowicz, *The War Against the Jews, 1933–1945* (New York: Bantam Books, 1986) p. 370.
3. According to Emilio Segrè, Fermi had special difficulties pronouncing the letter "r" in English. While he was in Ann Arbor teaching a summer course at the University of Michigan, a young American physicist volunteered to help him practice how to pronounce it by repeating the sentence: "Rear Admiral Byrd wrote a report concerning his travels in the southern part of the earth." Emilio

Segrè, *Enrico Fermi, Physicist* (Chicago: University of Chicago Press, 1970), p. 104.

4. Nobel Lectures, Physics 1922–1941 (Amsterdam, Netherlands: Elsevier Publishing, 1965).

CHAPTER 8

1. Otto Hahn to Lise Meitner, 19 December 1938, Meitner Collection, Churchill College Archives Centre, Cambridge, England. Barium is a metal similar to calcium, with atomic number 56 and atomic weight 137; it is not radioactive.
2. Hahn to Meitner, 19 December 1938, Meitner Collection.
3. Ruth Lewin Sime, *Lise Meitner: A Life in Physics* (Berkeley: University of California Press, 1996), p. 234.
4. Otto Frisch, *What Little I Remember* (New York: Cambridge University Press, 1979), p. 115.
5. Frisch, *What Little I Remember,* p. 115.
6. Frisch, *What Little I Remember,* p. 115.
7. One million electron volts, also equal to about 1.6 times 10^{-13} joules. It is the amount of energy gained by an electron accelerated in an electric field with potential difference of one volt, multiplied by a million. This amount of energy for the splitting of a single uranium nucleus agrees closely with what we know today.
8. Frisch, *What Little I Remember,* pp. 116–7.
9. Olga S. Opfell, *The Lady Laureates: Women Who Have Won the Nobel Prize* (Metuchen, NJ: Scarecrow Press, 1978), p. 109.

CHAPTER 9

1. AIP Oral History Interview of Werner Heisenberg, conducted by Thomas S. Kuhn, Max Planck Institute, Munich, November 30, 1962, Tape No. 39, Side 1.
2. Armin Hermann, *Werner Heisenberg, 1901–1976* (Bonn: Inter Nationes, 1976), p. 7.
3. AIP Oral History Interview of Werner Heisenberg, November 30, 1962, Tape No. 39, Side 1.
4. AIP Oral History Interview of Werner Heisenberg, November 30, 1962, Tape No. 39, Side 1.
5. Hermann, *Werner Heisenberg, 1901–1976,* p. 9.
6. AIP Oral History Interview of Werner Heisenberg, November 30, 1962, Tape No. 39, Side 1.
7. Hermann, *Werner Heisenberg, 1901–1976,* pp. 10–11.
8. Hermann, *Werner Heisenberg, 1901–1976,* p. 19.
9. Heisenberg's recollections, quoted in Hermann, *Werner Heisenberg, 1901–1976,* p. 19.
10. Hermann, *Werner Heisenberg, 1901–1976,* p. 33.
11. Letter from Heisenberg to Bohr, January 11, 1937, in Hermann, *Werner Heisenberg, 1901–1976,* p. 56.
12. Hermann, *Werner Heisenberg, 1901–1976,* p. 61.
13. Thomas Powers, *Heisenberg's War: The Secret History of the German Bomb* (New York: Knopf, 1993), p. 14.

CHAPTER 10

1. Spencer R. Weart and Gertrud Weiss Szilard, eds., *Leo Szilard: His Version of the Facts* (Cambridge: MIT Press, 1978), p. xvii.
2. Weart and Szilard, eds., *Leo Szilard: His Version of the Facts,* p. 3.
3. Weart and Szilard, eds., *Leo Szilard: His Version of the Facts,* pp. 15–17, and p. 17, n. 23.
4. Weart and Szilard, eds., *Leo Szilard: His Version of the Facts,* p. 17.
5. Weart and Szilard, eds., *Leo Szilard: His Version of the Facts,* p. 17.
6. Weart. and Szilard, eds., *Leo Szilard: His Version of the Facts,* p. 18, and p. 18, n. 28.
7. AIP Oral History Interview with Rudolph Ernst Peierls, conducted on June 17 and 18, 1963, by John L. Heilbron, American Institute of Physics, Center for the History of Physics, pp. 37–38.
8. Otto Frisch, *What Little I Remember* (New York: Cambridge University Press, 1979), p. 131.
9. Abraham Pais, *Niels Bohr's Times: In Physics, Philosophy, and Polity* (New York: Oxford University Press, 1991), p. 493.
10. Pais, *Niels Bohr's Times: In Physics, Philosophy, and Polity,* p. 494.
11. Pais, *Niels Bohr's Times: In Physics, Philosophy, and Polity,* p. 494.

CHAPTER 11

1. Thomas Powers, *Heisenberg's War: The Secret History of the German Bomb* (New York: Knopf, 1993), pp. 12–13.
2. Farm Hall, Operation Epsilon Transcripts, "Copy 1," at AIP, from the information on: July 14, and July 19, 1945, and revised after August 6, 1945; p. 11, and pp. XI and XII.
3. Powers, *Heisenberg's War,* p. 20.

CHAPTER 12

1. Thomas Powers, *Heisenberg's War: The Secret History of the German Bomb* (New York: Knopf, 1993), p. 121.
2. Robert Jungk, *Brighter Than a Thousand Suns: A Personal History of the Atomic Scientists.*(New York: Harcourt, Brace & World, 1958), p. 100.
3. Jungk, *Brighter Than a Thousand Suns,* pp. 100–1.
4. The reason for the title *The Virus House* is that the Nazis hid one of their nuclear research facilities, the one in the Kaiser Wilhelm Institute in Berlin-Dahlen, inside a building they named "The Virus House" in order to conceal what was happening inside and to deter anyone from wandering in. At that time fear of viruses was real, while few people knew much about the dangers of radiation. Powers, *Heisenberg's War: the Secret History of the German Bomb,* p. 121.
5. David Irving, *The Virus House: Germany's Atomic Research and Allied Counter-Measures.* London: Parforce UK Ltd., electronic version, 2002, p. 115.
6. Letter from Heisenberg to Robert Jungk, reprinted in Jungk, *Brighter Than a Thousand Suns,* pp. 102–4.
7. Powers, *Heisenberg's War: The Secret History of the German Bomb,* pp. 122–3.
8. Powers, *Heisenberg's War: The Secret History of the German Bomb,* p. 121.
9. These can be found at: http://nba.nbi.dk/papers/docs/cover.html.

10. This letter is reprinted by permission from the Niels Bohr Archive in Copenhagen. It is labeled Document 1, Translation, at http://nba.nbi.dk/papers/docs/d01tra.htm.

11. This letter is reprinted by permission from the Niels Bohr Archive in Copenhagen, labeled Document 11a, Translation, at http://nba.nbi.dk/papers/docs/d11atra.htm.

12. See Notes, chapter 11, in Powers, *Heisenberg's War: The Secret History of the German Bomb,* pp. 506–8.

13. Powers, *Heisenberg's War: The Secret History of the German Bomb,* p. 118.

14. From the Web site of the Haigerloch Museum in Germany: http://www.haigerloch.de/ceasy/modules/cms/main.php5?cPageId=95.

15. From the AIP, *Farm Hall Transcripts,* Operation Epsilon, August 8, 1945, also found at http://www.aip.org/history/heisenberg/p11a.htm.

16. AIP, *Farm Hall Transcripts,* Operation Epsilon, "Copy 1," at 7–13 September, 1945, Section I, par. 2.

17. Ominously, from my present-day vantage point, I remember that a surprisingly large percentage of the Iranian foreign students at the University of California at Berkeley (and at that time, the early 1970s, there were many foreign students from Iran in American universities) were studying nuclear engineering.

CHAPTER 13

1. Emilio Segrè, *Enrico Fermi, Physicist* (Chicago: University of Chicago Press, 1970), p. 120.

2. Segrè, *Enrico Fermi, Physicist,* p. 123.

3. Laura Fermi, *Atoms in the Family: My Life with Enrico Fermi* (Chicago: University of Chicago Press, 1954), p. 176.

4. Fermi, *Atoms in the Family,* p. 176.

5. Segrè, *Enrico Fermi, Physicist,* p. 123.

6. Segrè, *Enrico Fermi, Physicist,* p. 123.

7. Segrè, *Enrico Fermi, Physicist,* p. 124.

8. Segrè, *Enrico Fermi, Physicist,* p. 109.

9. Segrè, *Enrico Fermi, Physicist,* p. 128.

10. Segrè, *Enrico Fermi, Physicist,* p. 129.

11. AIP, Folder 23, Box 26, Samuel A. Goudsmit Papers, Niels Bohr Library, American Institute of Physics, College Park, MD.

CHAPTER 14

1. David Hawkins, Edith C. Truslow, and Ralph Carlisle Smith, *Project Y: The Los Alamos Story* (Los Angeles: Tomash Publishers, 1983), pp. 3–4.

2. Hawkins, Truslow, and Smith, *Project Y: The Los Alamos Story,* pp. 20–22.

3. Reminiscences of Norman F. Ramsey (July 19-August 4, 1960), conducted by Joan Stafford, p. 64, in the Columbia University Oral History Research Office Collection (hereafter CUOHROC). Quoted by permission of Columbia University.

4. Reminiscences of Norman F. Ramsey (July 19-August 4, 1960), pp. 64–5, CUOHROC.

5. Reminiscences of Norman F. Ramsey (July 19-August 4, 1960), p. 65, CUOHROC.

6. James W. Kunetka, *City of Fire: Los Alamos and the Birth of the Atomic Age, 1943–1945* (Englewood Cliffs, NJ: Prentice-Hall, 1978), p. 99.

7. Reminiscences of Norman F. Ramsey (July 19-August 4, 1960), p. 71, CUOHROC.

8. Reminiscences of Norman F. Ramsey (July 19-August 4, 1960), p. 67, CUOHROC.

9. Reminiscences of Norman F. Ramsey (July 19-August 4, 1960), p. 68, CUOHROC.

10. Reminiscences of Norman F. Ramsey (July 19-August 4, 1960), p. 68, CUOHROC.

11. Reminiscences of Norman F. Ramsey (July 19-August 4, 1960), p. 70, CUOHROC.

12. Reminiscences of Norman F. Ramsey (July 19-August 4, 1960), p. 80, CUOHROC.

13. Daniel J. Kevles, *The Physicists: The History of a Scientific Community in America* (New York: Knopf, 1978), p. 329.

14. Reminiscences of Norman F. Ramsey (July 19-August 4, 1960), p. 83, CUOHROC.

15. Reminiscences of Norman F. Ramsey (July 19-August 4, 1960), p. 84, CUOHROC.

16. "The Atomic Bomb and the End of World War II: A Collection of Primary Sources," National Security Archive Electronic Briefing Book No. 162, edited by William Burr, Document 2: Commander F. L. Ashworth to Major General L.R. Groves, "The Base of Operations of the 509th Composite Group," February 24, 1945, Top Secret. Source: RG 77, MED Records, Top Secret Documents, File no. 5g. The National Security Archive, George Washington University (Decoding provided by the Archive). These materials are reproduced from www.nsarchive.org with the permission of The National Security Archive.

17. "The Atomic Bomb and the End of World War II: A Collection of Primary Sources," National Security Archive Electronic Briefing Book No. 162, edited by William Burr, Document 10: General Lauris Norstad to Commanding General, XXI Bomber Command, "509th Composite Group; Special Functions," May 29, 1945, Top Secret. Source: RG 77, MED Records, Top Secret Documents, File no. 5g (copy from microfilm), National Security Archive, George Washington University (Decoding provided by the Archive). These materials are reproduced from www.nsarchive.org with the permission of The National Security Archive.

18. Reminiscences of Norman F. Ramsey (July 19-August 4, 1960), p. 121, CUOHROC.

19. The location of the test in the desert was selected around the end of 1944, and the name Trinity was chosen for this location by Robert Oppenheimer. He did not explain his choice, although he has said that he thought it had originated in poetry he had read, and perhaps was influenced by his reading of Indian mythology. In both cases, the "trinity" was derived from a trinity of deities.

20. Reminiscences of Norman F. Ramsey (July 19-August 4, 1960), p. 122, CUOHROC.

21. Kai Bird and Martin J. Sherwin, *American Prometheus: The Triumph and Tragedy of J. Robert Oppenheimer* (New York: Knopf, 2005), p. 309.

22. Reminiscences of Norman F. Ramsey (July 19-August 4, 1960), p. 124–125, CUOHROC.

23. Reminiscences of Norman F. Ramsey (July 19-August 4, 1960), p. 122–125, CUOHROC.

24. Reminiscences of Norman F. Ramsey (July 19-August 4, 1960), p. 126, CUOHROC.

25. Reminiscences of Norman F. Ramsey (July 19-August 4, 1960), p. 127, CUOHROC.

26. "The Atomic Bomb and the End of World War II: A Collection of Primary Sources," National Security Archive Electronic Briefing Book No. 162, edited by William Burr. Document 35: Cable War 33556 from Harrison to Secretary of War, July 17, 1945, Secret, Source: RG 77, MED Records, Top Secret Documents, File 5e (copy from microfilm), National Security Archive, George Washington University (Decoding provided by the Archive). These materials are reproduced from www.nsarchive.org with the permission of The National Security Archive.

27. "The Atomic Bomb and the End of World War II: A Collection of Primary Sources," National Security Archive Electronic Briefing Book No. 162, edited by William Burr. Document 36: Memorandum from General L. R. Groves to Secretary of War, "The Test," July 18, 1945, Top Secret, Excised Copy. Source: RG 77, MED Records, Top Secret Documents, File no. 4 (copy from microfilm), pp. 1–2, National Security Archive, George Washington University. These materials are reproduced from www.nsarchive.org with the permission of The National Security Archive.

28. "The Atomic Bomb and the End of World War II: A Collection of Primary Sources," National Security Archive Electronic Briefing Book No. 162, edited by William Burr, Document 36, p. 8.

CHAPTER 15

1. Eleanor Jette, *Inside Box 1663* (Los Alamos, NM: Los Alamos Historical Society, 1977), p. 109.

2. Abraham Pais, *Niels Bohr's Times: In Physics, Philosophy, and Polity* (New York: Oxford University Press, 1991), p. 492.

3. "The Atomic Bomb and the End of World War II: A Collection of Primary Sources," National Security Archive Electronic Briefing Book No. 162, edited by William Burr, Document 16: Memorandum from Arthur B. Compton to the Secretary of War, enclosing "Memorandum on 'Political and Social Problems,' from Members of the 'Metallurgical Laboratory' of the University of Chicago," June 12, 1945, Secret Source: RG 77, MED Records, H-B files, folder no. 76 (copy from microfilm), National Security Archive, George Washington University. These materials are reproduced from www.nsarchive.org with the permission of The National Security Archive.

4. Spencer R. Weart, and Gertrud Weiss Szilard, eds., *Leo Szilard: His Version of the Facts* (Cambridge: MIT Press, 1978), p. 181.

5. Weart and Szilard, eds., *Leo Szilard: His Version of the Facts*, pp. 181–2.

6. Weart and Szilard, eds., *Leo Szilard: His Version of the Facts*, p. 183.

7. Kai Bird and Martin J. Sherwin, *American Prometheus: The Triumph and Tragedy of J. Robert Oppenheimer* (New York: Knopf, 2005), p. 292.

8. Weart and Szilard, eds., *Leo Szilard: His Version of the Facts*, p. 183.

9. Weart and Szilard, eds., *Leo Szilard: His Version of the Facts*, pp. 187–8, and n.16, p. 188.

10. Television interview with Leo Szilard by Mike Wallace, 1961, AV 7–74–18, AIP, Niels Bohr Library and Archive.

11. Television interview with Leo Szilard by Mike Wallace.

12. AIP Oral History Interview with Norris Edwin Bradbury, conducted by Arthur Lawrence Norberg, February 11, 1976, p. 37, AIP, Niels Bohr Library and Archive. Quoted by permission of the Bancroft Library, University of California, Berkeley.

13. AIP Oral History Interview with Clyde Wiegand, conducted by Bruce Wheaton, September 26, 1977, pp. 26–7, AIP, Niels Bohr Library and Archive.

14. AIP Oral History Interview with Clyde Wiegand, p. 27.

15. AIP Oral History Interview with Clyde Wiegand, p. 28.

16. AIP Oral History Interview with Norris Edwin Bradbury, pp. 38–9.

17. AIP Oral History Interview with Norris Edwin Bradbury, p. 39.

18. AIP Oral History Interview with John H. Manley, conducted by Arthur Lawrence Norberg, July 9 and 11, 1976, pp. 53–4, AIP, Niels Bohr Library and Archive. Quoted by permission of the Bancroft Library, University of California, Berkeley.

19. Reminiscences of Norman F. Ramsey (July 19-August 4, 1960), conducted by Joan Stafford, pp. 72–74, in the Columbia University of Oral History Research Office Collection (hereafter CUOHROC). Quoted by permission of Columbia University.

20. Reminiscences of Norman F. Ramsey (July 19-August 4, 1960), p. 72, CUOHROC.

21. Reminiscences of Norman F. Ramsey (July 19-August 4, 1960), p. 73, CUOHROC.

22. Reminiscences of Norman F. Ramsey (July 19-August 4, 1960), pp. 75–76, CUOHROC.

23. Reminiscences of Norman F. Ramsey (July 19-August 4, 1960), p. 76, CUOHROC.

24. AIP Oral History Interview with Robert Serber, conducted by Charles Weiner and Gloria Lubkin, February 10, 1967, p. 50, AIP, Niels Bohr Library and Archive.

25. AIP Oral History Interview with Robert Serber, pp. 50–1.

26. AIP Oral History Interview with Robert Serber, pp. 51–2.

27. Robert P. Newman, *Truman and the Hiroshima Cult* (East Lansing, MI: Michigan State University Press, 1995), p. 103.

28. Newman, *Truman and the Hiroshima Cult,* p. 105.

29. Peter Wyden, *Day One* (New York: Simon & Schuster, 1984), p. 15.

CHAPTER 16

1. "The Atomic Bomb and the End of World War II: A Collection of Primary Sources," National Security Archive Electronic Briefing Book No. 162, edited by William Burr, Document 3a: Memorandum Discussed with the President, April 25, 1945, Source: Henry Stimson Diary, Manuscripts and Archives, Yale University Library, Henry Lewis Stimson Papers (microfilm at Library of Congress). These materials are reproduced from www.nsarchive.org with the permission of The National Security Archive.

2. "The Atomic Bomb and the End of World War II: A Collection of Primary Sources," National Security Archive Electronic Briefing Book No. 162, edited by William Burr. See the description of Documents 3a-d, National Security Archive, George Washington University, at: www.nsarchive.org.

3. "The Atomic Bomb and the End of World War II: A Collection of Primary Sources," National Security Archive Electronic Briefing Book No. 162, edited

by William Burr, Document 3c, untitled memorandum by General L.R. Groves, April 25, 1945, Source: Record Group 200, Papers of General Leslie R. Groves, Correspondence 1941–1970, box 3, "F," National Security Archive, George Washington University. These materials are reproduced from www.nsarchive.org with the permission of The National Security Archive.

4. "The Atomic Bomb and the End of World War II: A Collection of Primary Sources," National Security Archive Electronic Briefing Book No. 162, edited by William Burr, Document 3b: Memorandum for the Secretary of War from General L. R. Groves, "Atomic Fission Bombs," April 23, 1945, Source: RG 77, Commanding General's file no. 24, tab D, National Security Archive, George Washington University. These materials are reproduced from www .nsarchive.org with the permission of The National Security Archive.

5. "The Atomic Bomb and the End of World War II: A Collection of Primary Sources," National Security Archive Electronic Briefing Book No. 162, edited by William Burr, Document 7: Diary Entries, May 14 and 15, 1945, Source: Henry Stimson Diary, Manuscripts and Archives, Yale University Library, Henry Lewis Stimson Papers (microfilm at Library of Congress). These materials are reproduced from www.nsarchive.org with the permission of The National Security Archive.

6. "The Atomic Bomb and the End of World War II: A Collection of Primary Sources," National Security Archive Electronic Briefing Book No. 162, edited by William Burr, Document 15: Memorandum of Conference with the President, June 6, 1945, Top Secret. Source: Manuscripts and Archives, Yale University Library, Henry Lewis Stimson Papers (microfilm at Library of Congress). These materials are reproduced from www.nsarchive.org with the permission of The National Security Archive.

7. "The Atomic Bomb and the End of World War II: A Collection of Primary Sources," National Security Archive Electronic Briefing Book No. 162, edited by William Burr, Document 29: "Magic"–Diplomatic Summary, War Department, Office of Assistant Chief of Staff, G-2, No. 1204–July 12, 1945, Top Secret Ultra, Source: Record Group 457, Records of the National Security Agency/Central Security Service, "Magic" Diplomatic Summaries 1942–1945, box 18, National Security Archive, George Washington University. These materials are reproduced from www.nsarchive.org with the permission of The National Security Archive.

8. "The Atomic Bomb and the End of World War II: A Collection of Primary Sources," National Security Archive Electronic Briefing Book No. 162, edited by William Burr, Document 30: John Weckerling, Deputy Assistant Chief of Staff, G-2, July 12, 1945, to Deputy Chief of Staff, "Japanese Peace Offer," 13 July 1945, Top Secret Ultra, Source: RG 165, Army Operations OPD Executive File #17, Item 13 (copy courtesy of J. Samuel Walker), National Security Archives, George Washington University. These materials are reproduced from www.nsarchive.org with the permission of The National Security Archive.

9. "The Atomic Bomb and the End of World War II: A Collection of Primary Sources," National Security Archive Electronic Briefing Book No. 162, edited by William Burr, Document 31: "Magic"–Diplomatic Summary, War Department, Office of Assistant Chief of Staff, G-2, No. 1205–July 13, 1945, Top Secret Ultra, Source: Record Group 457, Records of the National Security Agency/Central Security Service, "Magic" Diplomatic Summaries 1942–1945, box 18, National Security Archive, George Washington University. These ma-

terials are reproduced from www.nsarchive.org with the permission of The National Security Archive.

10. "The Atomic Bomb and the End of World War II: A Collection of Primary Sources," National Security Archive Electronic Briefing Book No. 162, edited by William Burr, Document 40, p. 2, bottom. "Magic"–Diplomatic Summary, War Department, Office of Assistant Chief of Staff, G-2, No. 1214–July 22, 1945, Top Secret Ultra, Source: Record Group 457, Records of the National Security Agency/Central Security Service, "Magic" Diplomatic Summaries 1942–1945, box 18, National Security Archive, George Washington University. These materials are reproduced from www.nsarchive.org with the permission of The National Security Archive.

11. "The Atomic Bomb and the End of World War II: A Collection of Primary Sources," National Security Archive Electronic Briefing Book No. 162, edited by William Burr, Document 44: "Magic"–Diplomatic Summary, War Department, Office of Assistant Chief of Staff, G-2, No. 1221–July 29, 1945, Top Secret Ultra, Source: Record Group 457, Records of the National Security Agency/Central Security Service, "Magic" Diplomatic Summaries 1942–1945, box 18, National Security Archive, George Washington University. These materials are reproduced from www.nsarchive.org with the permission of The National Security Archive.

12. "The Atomic Bomb and the End of World War II: A Collection of Primary Sources," National Security Archive Electronic Briefing Book No. 162, edited by William Burr, Document 37: Diary Entry for July 20, 1945. Source: Takashi Itoh, ed., *Sokichi Takagi: Nikki to Joho [Sokichi Takagi: Diary and Documents]* (Tokyo, Japan: Misuzu-Shobo, 2000), 916–917 [trans. Hikaru Tajima], National Security Archive, George Washington University. These materials are reproduced from www.nsarchive.org with the permission of The National Security Archive.

CHAPTER 17

1. "The Atomic Bomb and the End of World War II: A Collection of Primary Sources," National Security Archive Electronic Briefing Book No. 162, edited by William Burr, Document 1: Memorandum from Vannevar Bush and James B. Conant, Office of Scientific Research and Development, to Secretary of War, September 30, 1944, Top Secret, Source: Record Group 77, Records of the Army Corps of Engineers (hereinafter RG 77), Manhattan Engineering District (MED), Harrison-Bundy Files (H-B Files), folder 69, National Security Archive, George Washington University. These materials are reproduced from www.nsarchive.org with the permission of The National Security Archive.

2. "The Atomic Bomb and the End of World War II: A Collection of Primary Sources," National Security Archive Electronic Briefing Book No. 162, edited by William Burr, Document 1. These materials are reproduced from www.nsarchive.org with the permission of The National Security Archive.

3. Helen S. Hawkins, G. Allen Gerb, and Gertrud Weiss Szilard, eds., *Toward a Livable World: Leo Szilard and the Crusade for Nuclear Arms Control* (Cambridge: MIT Press, 1987), p. 253.

4. "The Nation's Future," NBC Television, November 12, 1960, excerpted in Hawkins, Gerb, and Szilard, eds., *Toward a Livable World,* p. 239.

5. Frank L Gertcher and William J. Weida, *Beyond Deterrence: The Political Economy of Nuclear Weapons* (Boulder: Westview Press, 1990), p. 19.

6. Freeman Dyson, *Weapons and Hope* (New York: Harper & Row, 1984), p. 240.
7. David C. Hendrickson, *The Future of American Strategy* (New York: Holmes & Meier, 1987), p. 117.
8. Walter Lippman, *The Cold War: A Study of U.S. Foreign Policy* (New York: Harper & Row, 1947), p. 119.
9. Henry A. Kissinger, *Nuclear Weapons and Foreign Policy* (New York: Doubleday, 1957), p. 189.
10. Bart Brasher, *Implosion: Downsizing the U.S. Military, 1987–2015.* (Westport, CT: Greenwood Press, 2000), p. 105.
11. Brasher, *Implosion: Downsizing the U.S. Military,* p. 105.

CHAPTER 18

1. Hans Binnendijk and Richard L. Kugler, *Seeing the Elephant: The U.S. Role in Global Security* (Dulles, VA: Potomac Books, 2006), p. 162.
2. Adam Liptak, "Justices Back Navy on Sonar Use," *New York Times,* November 13, 2008.

REFERENCES

Note: All references in this book designated AIP refer to material from the Niels Bohr Library and Archive of the American Institute of Physics, College Park, Maryland.

Asimov, Isaac. *Inside the Atom.* New York: Abelard-Schuman, 1966, 3rd ed.

Bernstein, Jeremy. *Hitler's Uranium Club: The Secret Recordings at Farm Hall.* Intro. David Cassidy. Woodbury, NY: American Institute of Physics, 1996.

Binnendijk, Hans, and Richard L. Kugler. *Seeing the Elephant: The U.S. Role in Global Security.* Dulles, VA: Potomac Books, 2006.

Bird, Kai, and Martin J. Sherwin, *American Prometheus: The Triumph and Tragedy of J. Robert Oppenheimer.* New York: Knopf, 2005.

Brasher, Bart. *Implosion: Downsizing the U.S. Military, 1987–2015.* Westport, CT: Greenwood Press, 2000.

Broda, Engelbert. *Ludwig Boltzmann: Man, Physicist, Philosopher.* Woodbridge, CT: Ox Bow Press, 1983.

Chandler, Robert W., and John R. Backschies. *The New Face of War.* McLean, VA: Amcoda Press, 1999.

Cropper, William H. *Great Physicists: The Life and Times of Leading Physicists from Galileo to Hawking.* New York: Oxford University Press, 2001.

Dyson, Freeman. *Weapons and Hope.* New York: Harper & Row, 1984.

Farm Hall, Operation Epsilon Transcripts, "Copy 1," at AIP.

Fermi, Laura. *Atoms in the Family: My Life with Enrico Fermi.* Chicago: University of Chicago Press, 1954.

Frayn, Michael. *Copenhagen.* New York: Anchor Books, 1998.

Frayn, Michael, and David Burke. *The Copenhagen Papers: An Intrigue.* New York: Metropolitan Books, 2000.

Frisch, Otto. *What Little I Remember.* New York: Cambridge University Press, 1979.

Gamow, George. *Thirty Years that Shook Physics: The Story of Quantum Theory.* New York: Doubleday, 1966.

Gertcher, Frank L., and William J. Weida. *Beyond Deterrence: The Political Economy of Nuclear Weapons.* Boulder: Westview Press, 1990.

Hahn, Otto. *My Life.* London: Macdonald, 1968.

Hawkins, David, Edith C. Truslow, and Ralph Carlisle Smith. *Project Y: The Los Alamos Story.* Los Angeles, CA: Tomash Publishers, 1983.

Hawkins, Helen S., G. Allen Gerb, and Gertrud Weiss Szilard, eds. *Toward a Livable World: Leo Szilard and the Crusade for Nuclear Arms Control.* Cambridge: MIT Press, 1987.

Hendrickson, David C. *The Future of American Strategy.* New York: Holmes & Meier, 1987.

Herken, Gregg. *Brotherhood of the Bomb: The Tangled Lives and Loyalties of Robert Oppenheimer, Ernest Lawrence, and Edward Teller.* New York: Henry Holt, 2002.

Hermann, Armin. *Werner Heisenberg, 1901–1976.* Bonn: Inter Nationes, 1976.

Huntington, Samuel P. *Clash of Civilizations and the Remaking of World Order.* New York: Simon & Schuster, 1996.

Irving, David. *The Virus House.* London: Parforce, electronic edition, 2002; 1967.

Jette, Eleanor. *Inside Box 1663.* Los Alamos, NM: Los Alamos Historical Society, 1977.

Jungk, Robert. *Brighter than a Thousand Suns: A Personal History of the Atomic Scientists.* New York: Harcourt, Brace & World, 1958.

Kanda, Mikio, ed. *Widows of Hiroshima: The Life Stories of Nineteen Peasant Wives.* Trans. Taeko Midorikawa. New York: St. Martin's Press, 1989.

Kevles, Daniel J. *The Physicists: The History of a Scientific Community in America.* New York: Knopf, 1978.

Kissinger, Henry A. *Nuclear Weapons and Foreign Policy.* New York: Doubleday, 1957.

Kunetka, James W. *City of Fire: Los Alamos and the Birth of the Atomic Age, 1943–1945.* Englewood Cliffs, NJ: Prentice-Hall, 1978.

Lippman, Walter. *The Cold War: A Study of U.S. Foreign Policy.* New York: Harper & Row, 1947.

Liptak, Adam. "Justices Back Navy on Sonar Use," *New York Times,* November 13, 2008.

Maddox, Robert James. *Weapons for Victory: The Hiroshima Decision.* Columbia: University of Missouri Press, 2004.

Meitner, Lise. Letters and papers in Meitner Collection, Churchill College Archives Centre, Cambridge, England.

Moore, Ruth. *Niels Bohr: The Man, His Science, and the World They Change*d. Cambridge: MIT Press, 1985.

Newman, Robert P. *Truman and the Hiroshima Cult.* East Lansing, MI: Michigan State University Press, 1995.

Oe, Kenzaburo. *Hiroshima Notes.* New York: Marion Boyars, 1995.

Opfell, Olga S. *The Lady Laureates: Women Who Have Won the Nobel Prize.* Metuchen, NJ: Scarecrow Press, 1978.

Pais, Abraham. *Niels Bohr's Times: In Physics, Philosophy, and Polity.* New York: Oxford University Press, 1991.

Pais, Abraham. *J. Robert Oppenheimer: A Life.* New York: Oxford University Press, 2006.

Powers, Thomas. *Heisenberg's War: The Secret History of the German Bomb.* New York: Knopf, 1993.

Rhodes, Richard. *The Making of the Atomic Bomb.* New York: Simon & Schuster, 1986.

Rozental, S., ed. *Niels Bohr: His Life and Work as Seen by His Friends and Colleagues.* New York: Wiley, 1967.

Segrè, Emilio. *Enrico Fermi, Physicist.* Chicago: University of Chicago Press, 1970.

Segrè, Emilio. *From X-Rays to Quarks: Modern Physicists and Their Discoveries.* Berkeley: University of California Press, 1980.

Sime, Ruth Lewin. *Lise Meitner: A Life in Physics.* Berkeley: University of California Press, 1996.

Sylves, Richard T. *The Nuclear Oracles: A Political History of the General Advisory Committee of the Atomic Energy Commission, 1947–1977.* Ames: Iowa State University Press, 1987.

Thorpe, Charles. *Oppenheimer: The Tragic Intellect.* Chicago: University of Chicago Press, 2006.

Trefil, James S. *From Atoms to Quarks: An Introduction to the Strange World of Particle Physics.* New York: Scribner, 1979.

Weart, Spencer R., and Gertrud Weiss Szilard, eds. *Leo Szilard: His Version of the Facts.* Cambridge: MIT Press, 1978.

Wyden, Peter. *Day One.* New York: Simon & Schuster, 1984.

INDEX

Advance Praise for *Learning Sickness*

"Thank you, Jim Lang, for sharing such a heartfelt account of one man's journey through illness and back to health. Honestly, I couldn't put it down! The insights in these pages, so wonderfully written in vivid detail, bring me back through my own early trials with Crohn's disease. This is the perfect book for all people tackling chronic illness who feel they may be alone."

—Jill Sklar, author of *The First Year–Crohn's Disease and Ulcerative Colitis* and *Eating for Acid Reflux*

"In his book *Learning Sickness*, James Lang ignites an incandescent passage to understanding chronic illness. His personal path is strewn with rocks but his spirit refuses to be turned away from learning to deal with his secret pain. Sometimes truth assumes the appearance of a bitter pill, but Lang looks to a greater message when he observes 'life pulses on.' He describes a journey that is in part a search for a miracle cure, but is actually an amazing discovery of a comfort zone that waited for him all along. The reader will never hear of the 'glue' in relationships or hold the hand of a child again without a reminder of James Lang and his wisdom to look at chronic illness in its full measure and influence."

—Barbara Wolcott, Pulitzer Prize nominee and author of *David, Goliath and the Beach-Cleaning Machine*

"As a patient with Crohn's disease, I cannot help but be awed by the bravery that is so evident not only in Jim Lang's struggle to accept his illness, but also in his willingness to share his experiences. As a legal advocate for other patients with Crohn's disease, I see Jim's book as a critically important tool that will help others—doctors, insurers, government officials, and others who control the flow of resources to patients—to understand what we live with day-to-day. The worst thing about having a chronic illness is how isolating it is. *Learning Sickness* is a bridge to overcoming that isolation."

—Jennifer Jaff, attorney

"Written with clarity and honesty, James Lang's chronicle of his illness is highly informative. But Lang is intent on a higher goal as well—a chronicle of inner change and of the individual's development of a vital kind of human freedom that can discover a humble, daily but ultimately priceless meaning in life, despite—or even through—suffering."

—Reginald Gibbons, Professor and Chair, Department of English, Northwestern University, author of *Sweetbitter*

"A touching and true account of what life can be like with Crohn's disease and the impact it has on one's life . . . ."

—Annette D'Ercole, fellow inflammatory bowel disease (IBD) sufferer and mother of a child with Crohn's disease

"One of Lang's central lessons is that suffering from a chronic illness means learning to recognize yourself as a person larger than your illness. In the same way, his book is larger than a meditation on Crohn's disease; it's a manual on living a richer life.

This book explores the paradox that meditating on sickness can lead us to recognize what true health is."
—Joe Kraus, author of *The Accidental Anarchist*

" 'The human body,' James Lang writes, 'does not present itself . . . as an open textbook. It is instead, like everything we find in the natural world: complex, mysterious, and capable of bursting through the categories of knowledge we have constructed to contain it.' This 'bursting through' is just what happens when Lang's own body begins to instruct him, forcing him to revise his life. *Learning Sickness* is a book about being educated in the difficult school of illness, and it points to the hard-won wisdom that may be found there."
—Mark Doty, author of *Heaven's Coast* and *Firebird*

"Chronic illness in general, and inflammatory bowel disease in particular, have found a voice in Jim Lang's *Learning Sickness*. It is a voice of wisdom, of courage, of hope, and of inspiration. You will be touched and inspired by a powerful story told by a gifted writer."
—Marvin Bush, managing partner of Winston Partners

"*Learning Sickness* is a beautifully written, frank account of one man's struggle to come to terms with the chronic illness that altered every aspect of his life. This is an inspiring and informative book that will be of value to all patients, whether they have lived with inflammatory bowel disease for many years or have just been diagnosed."
—Rodger DeRose, President and CEO, Crohn's and Colitis Foundation of America (CCFA)

"An honest and reflective first person narrative of the journey of a young professional to self-awareness and appreciation of life after being diagnosed with Crohn's disease. Written in the spirit of Anatole Broyard's *Intoxicated by My Illness*, but instead of writing about dying as an end-point, Lang writes about learning how to live each day fully. Even those who are not living with chronic illness will benefit from Lang's struggle to live each and every day. Highly recommended for medical practitioners and laypeople alike."
—Jeanette J. Norden, Ph.D., Professor and Director of Medical Education, Department of Cell and Developmental Biology, Vanderbilt University Medical School

"Lang's work is more than a brilliant description of illness; it is an articulation of health, family, and recovery. Honest, insightful, often funny, this is a book that makes you appreciate the sheer animal pleasure of being alive. Read it and experience a strong new voice."
—David Gessner, author of *Return of the Osprey*

"Eloquent and honest, this book is not merely the story of a man negotiating the specific obstacles posed by Crohn's disease and the medical establishment: It is ultimately the tale of a spiritual journey, shaped by the author's commitment to come to terms with the hard lessons taught by chronic disease."
—Michael Land, assistant professor of English, Assumption College

LEARNING SICKNESS

From *Capital Discoveries*, books that offer journeys of self-discovery, transformation, inner-awareness, and recovery. Other titles include:

Cancer Happens
by Rebecca Gifford

My Renaissance
by Rose Marie Curteman

There's a Porcupine in My Outhouse
by Michael J. Tougias

Reading Water
by Rebecca Lawton

Rivers of a Wounded Heart
by Michael Wilbur

Ancient Wisdom
by P. Zbar

Daughters of Absence
by Mindy Weisel

Wise Women Speak to the Woman Turning 30
by Jean Aziz and Peggy Stout

Swimming with Maya
by Eleanor Vincent

LEARNING SICKNESS

A Year with Crohn's Disease

James M. Lang

A CAPITAL DISCOVERIES BOOK

CAPITAL
BOOKS, INC.
Sterling, Virginia

Capital Books, Inc.
P.O. Box 605
Herndon, Virginia 20172-0605

ISBN 1-931868-60-3 (alk.paper)

Library of Congress Cataloging-in-Publication Data

Lang, James M.
   Learning sickness : a year with Crohn's disease / James M. Lang.—1st ed.
      p.   cm.
   Includes bibliographical references and index.
   ISBN 1-931868-60-3
   1. Lang, James M.—Health.   2. Crohn's disease—Patients—United States—Biography.   I. Title.
   RC862.E52L36   2003
   362.1'963445'0092—dc22   2003014801

To Anne

I suppose what I'm really to render is nothing more nor less than Life—as one man has found it.

H.G. Wells, *Tono-Bungay*

Contents

Acknowledgments

❧

Of course I have many debts to acknowledge in the writing of this book.

As each chapter came fresh off the printer, it went immediately to Mike Land, who read them all carefully and thoughtfully, and provided feedback and suggestions on many levels. Once the book was complete, David Thoreen helped me refine the language with a precision and attention to words that only a poet could offer. Joe Kraus read an early version of the book, and offered several major suggestions that ultimately made the book a far more coherent piece of work that it would otherwise have been. Rachel Ramsey and Mike O'Shea read completed drafts and offered support and encouragement that I was much in need of at the time. Jeanette Norden offered me invaluable information and advice about medicine, psychology, and writing.

My siblings and in-laws all read the book in draft form as well, and to each of them I am grateful for their enthusiasm and encouragement. I have to make a special acknowledgement to Jon Roketenetz for his design of the Web site we created for the book while I was writing it, and for inspiring me with his own artistic vision and commitment. I must make an equally special acknowledgment to my brother Tony, whom you will meet in the pages that follow, and who has been—for all of my thirty-three years on this planet—a mentor and friend, as well as an older brother.

My mother and father have been and still are constant sources of support and love, and I am eternally grateful for what they provided for me on every level of existence: physical, emotional, intellectual, spiritual. Together they have modeled for me what it means to live well.

How Anne, Katie, Madeleine—and now Jillian, who is turning one year old on the day I am writing this sentence—contributed to the making of this book will be clear enough in the narrative itself, so to them I simply offer again my love and gratitude for all they have given and taught to me.

I am grateful as well to Drs. Robert Honig and John Darrah for their care of me during the year described in this book.

Finally, I want to acknowledge and thank my agent, Sandy Choron, and editor, Noemi Taylor, whose suggestions, praise, and enthusiasm for the book made me a believer in my own work.

Preface

~

This book follows the course of a single year in my life, a year in which I struggled through a long and active phase of a chronic illness that causes inflammation, ulceration, and hemorrhaging in the lining of the digestive tract. Crohn's disease is one of many chronic illnesses characterized by the maddening habit of alternating between phases of sustained disease activity, which may last for weeks or months or even years, and phases of total remission, in which individuals may live for weeks or months or even years without any symptoms whatsoever. The maddening part is that no apparent rhyme nor reason dictates what sends the disease from one phase to the next.

For most of the year described in this book, the disease was active—more active than it had ever been in my half-dozen years of living with it, and active for the longest uninterrupted stretch I had experienced thus far. In that respect, this year represented an unusual one in the history of my illness. For some time prior to the year narrated in the book, the disease had been in remission; for six months following that year, the disease returned to a state of complete remission.

I should note that my specific Crohn's disease story, like any specific Crohn's disease story, should not be taken as a typical or general account of the disease. Every case, like every human being, is unique. Some fortunate sufferers will never experience the sort of exacerbation of symptoms that led to the hospitalization I describe in chapter six; some unfortunate sufferers will never know the joys of the remission I describe in chapters nine and ten. A majority of sufferers will have to deal with surgery at some point in their lives, a treatment option to which I have not yet had to resort.

But I am confident that most readers with Crohn's disease or its companion illness, ulcerative colitis, even if their physical symptoms and disease history vary from mine, will be able to identify with the emotional, intellectual, and spiritual reactions I had to the disease's attacks on my body. And I am hopeful that anyone who has ever confronted the limitations and defects of a physical human body, even temporary ones, will be able to identify with the lessons I learned from my disease.

Chronologically, my story in this book begins and ends in remission. In the opening pages of the first chapter I describe the final days of my life in remission, in July and early August of 2000; in the final two chapters I describe the slow return of the disease to remission, in July and August of 2001.

What I hope to convey is the evolution I underwent, as a human being, in the time spent between those two remissions. I tried to capture the essence of that evolution in the lessons I describe in each chapter. Those lessons begin, as my evolution began, with a focus on the nature of chronic illness, on disease, and upon the ways we respond to disease both as individuals and as a society. But they gradually become broader in focus, and move into more complex and universal human territories.

When I originally conceived this project, in December of 2000, I envisioned it as sociological study in which I would interview individuals with chronic illness and capture the insights and wisdom that I believed disease could offer to those of us who suffered from it. I firmly believed then—and I retain this belief still—that illness and suffering can bring a wisdom that remains unavailable to those who have not yet—and may never—have experienced them. I believe that illness and suffering can offer us insights into the most important questions we can ask ourselves about what it means to be a human being: Why do we exist? How should we behave toward one another? How do we love our spouses, and raise our children, and live our daily lives in meaningful and creative ways? I wanted this book to compile and present the collective wisdom of the chronically ill in response to these fundamental questions.

In preparation for writing that book, I sat down to tell my own story, and to think about the particular wisdom that I felt disease had brought to me. How had my responses to these questions been transformed over a year of illness and suffering? I began to write my story, and that story grew, and grew, and grew, until eventually I realized that I had all I could handle in telling my own story.

I still hope one day to write that other book, but this is the book that came out of my original intention. I will be interested to learn whether others who have undergone similar experiences with chronic illness have come to the same conclusions I have.

One final note about the book's organization.

I have tried to pin each chapter to a specific portion of the year described in the book, and those sections are arranged in chronological order—with the exception of chapter two, which flashes back to earlier periods in my life in order to provide some background on how I first learned to think about chronic illness and the human body. The dates at the beginning of each chapter identify roughly the month or season narrated in that chapter.

However, almost all of the chapters also involve movement in and out of the chronological moment identified in that date—sometimes to reflect back upon moments that had led me to the chronological moment, and sometimes to look forward to moments when incidents described in that chapter come to a close later in the chronological year, or even beyond the year described in the book. The real connecting themes of each chapter are the lessons. The time period that I identify for each chapter, then, really pinpoints the time period in which I felt that I learned that particular lesson.

I allude to each of those lessons in the chapter titles. But they are not—with the exception of the epilogue, which is a private note to my fellow Crohn's disease sufferers—offered in bulleted or numbered lists within the chapter. To offer them in that format would have made them too easily packaged, too divorced from the experiences in which they were learned, and hence too easy to ignore or file away with all of the other bits of wisdom and advice we get from our friends and our families, from the television, from our pastors, from the comic pages, and from the latest self-help craze.

I hope that the reader will see and understand these lessons as I understood them—by living through them, in each chapter, along with me.

1

LEARNING TO BE ILL

❧

[July–August 2000]

For the record, it was my wife's idea to take in a show at the transvestite club on Bourbon Street in New Orleans. She had been trying to pull me into it all weekend, and Saturday night she broke down my resistance. Earlier that summer she had been to a transvestite show in Chicago, which had featured good-looking men in sequined gowns singing campy show tunes and 1980s hits, and she and her friends had loved it. It didn't sound half-bad for an evening of entertainment—except for the part about the show tunes—so I let her drag me into a dark and very seedy bar early that evening.

That Saturday night was our third and final evening of vacation, capping off a four-day trip to New Orleans in late July of 2000 by ourselves—no children, no other couples, no work or obligations of any sort. We were staying in a small hotel a block off Bourbon Street, and had developed a pattern since our arrival on Wednesday evening. In the mornings we slept in until noon or so, and then

1

ventured out for lunch in the Quarter. In the afternoons we would see some sights—take a river cruise, or visit the aquarium. In the evenings we had dinner out, and from dinner until at least midnight or so we would stroll up and down Bourbon Street, drinking "to go" cups of beer or rum hurricanes and wandering in and out of the bars, listening to music and watching the crowds.

More of such aimless wandering would have easily fulfilled my ambitions that Saturday night, but Anne had this better idea in mind—or an idea that seemed better to us at ten o'clock that Saturday night.

Our first clue should have been the individual outside hawking the show. He was no glamorous transvestite outfitted in a beautifully sequined gown. He had the requisite long hair and fake breasts, but he also had about fifty extra pounds for his frame, and they were shoved into a dress that belonged on a man half his size. As he winked and greeted us, I hoped that there would be no stripping.

There wasn't—there wasn't much of anything but bad singing, clumsy dancing, and the inescapable stench of male sweat mingled with perfume. The waiters were all in drag, and were very insistent about the club's two-drink minimum. In the bathroom I heard two transvestites cattily running down the outfit of a third.

The whole affair was sordid enough to be depressing, but as we stumbled out of the mercifully short show we couldn't help but laugh: that had been one *lousy* transvestite show. The bad taste in our mouths was nothing that wouldn't be cured by a few hours of strolling up and down Bourbon Street, drinking beer and listening to live music—which is how we finished the evening, and our trip to New Orleans.

It had been an absolutely idyllic vacation for us in a number of ways. First and foremost, we were on vacation alone together for the first time since our honeymoon. Even that honeymoon hadn't counted as much of a vacation. We spent seven cold and rainy January days in a condominium in Florida loaned to us by my parents' best friends, and for the entire time my wife had a terrible cold.

From our honeymoon through the birth of our first child, a period of around three years, our vacations consisted exclusively of summertime visits to our parents' homes, and trips to friends' weddings. For both sorts of trips, we were always among crowds. After our children started coming, in December of 1995, vacations on our own were no longer an option. Even when we could get family members to watch our children overnight or for a weekend, it was invariably for some event involving other people.

It took us until the summer of 2000 to con my younger sister into watching our two children for a span of four days, so we could at last have a vacation by ourselves.

As terrible as this may sound, the chance to be alone together with Anne reminded me how much fun we could have without our children, and I vowed afterward that we would continue to take vacations away from them. Despite all of the joy they can provide, children can be very difficult on vacations. Carsick, hungry, thirsty, complaining, wet-diapered, bored, tired, anxious, irritable, and inflexible—these were the adjectives that sprang to mind when I thought of car trips with our children, then ages three and five.

But the time we had alone together did not account for all of the magic of that trip. Part of it, no doubt, was New Orleans itself.

In the afternoons we usually found time to stop at the famous Café du Monde in Jackson Square, where we would drink coffee and have a beignet—a delicate French pastry absolutely smothered in powdered sugar. From our seat in the café we could watch the pedestrian traffic streaming around Jackson Square, and just across the street we could see the palm readers plying their trade at cardboard tables up and down the walkways.

One afternoon we took a paddleboat cruise down the Mississippi River, learning about the history of New Orleans and its surrounding areas from a guide as we went. We followed his monologue for most of the way, but we spent some time in the dining room as well, having a few drinks and listening to the live band they had on board. Eventually we took our drinks outside and sat near the railings on the boat's side, watching the water roll by and enjoying the gentle breeze.

Of course it was New Orleans, so the meals were a major part of the experience, and a major part of each day. We planned two touristy dinners—one at Emeril's, and one at the Commander's Palace, a restaurant that had been repeatedly ranked number one in the country by a major tour guide publication. The rest of the time we simply stopped in whatever restaurant caught our fancy in our wanderings in the Quarter. We ate plenty of Cajun food, drank plenty of beer and rum, and didn't have a bad meal all week.

One evening we settled down into a small bar to listen to a band playing Cajun music, which included a washboard as one of its instruments. I was fascinated by this instrument, the playing of which consisted of running a spoon rhythmically up and down across the slats of the metal washboard. Another evening we climbed up into the balcony of a bar overlooking the street, and stood and watched people walking by as we sipped our beers and leaned on the railing, joined by dozens of college students.

Throughout our evening strolls we would go in and out of the little shops on Bourbon Street—the voodoo shop way up at the far end, with its impossible-to-follow injunctions to take the religion of the patrons seriously; the sex shop down at the other end, where we bought a few joke gifts for friends and family; and the various junk shops along the way, where we bought feather boas and beads for our daughters to add to their dress-up collections.

And throughout all of this—in the noisy Café du Monde, over the white tablecloths of the Commander's Palace, in the gently rocking dance room of the paddleboat, and in and out of the crowds on Bourbon Street—we talked. In those four days we probably talked more than we ever had as a couple. We talked about how our lives were going thus far, whether they were turning out as we had imagined them, and what they might have in store for us in the future. Anne, an elementary-school teacher, had begun to think about a new career in health care—perhaps even returning to school for a medical degree. I was beginning a new job in a month, and had plenty of anxiety and excitement to share with her about that. We realized together how much we enjoyed traveling, and imagined other trips we might take. Our conversations were infused with a sense of hope and anticipation for the future.

"It's exciting," I remember saying to her one evening over dinner. "How wide open the future can be."

All in all, it was an incredible vacation, and by the time our plane landed back in Chicago we were already fantasizing about a return trip.

In retrospect, I suspect I may see this vacation as especially idyllic because it marked the final days of health I experienced before my yearlong descent into the heart of chronic illness. I knew I was enjoying myself while I was in New Orleans; as the year I will describe in this book wore on, I began gradually to see that time in New Orleans as the moment in which I was dancing on the cliff edge, oblivious to the tumble I was about to take. The subsequent knowledge of what I found at the bottom of the cliff slowly transformed that dance into an ever more magical and perfect time of innocence for me.

But New Orleans was not just a distinctive moment in the history of the health of my physical body; it marked, too, the final moments before I was forced to undergo some profound intellectual, emotional, and spiritual transformations. I see New Orleans now as the very last time in my life when I had been able simply to let go—to abandon thoughts of my illness, of my diet, of sobriety, of the beer and spicy food I was putting into my system, and simply enjoy myself for four days.

In New Orleans I ate and drank as I had been eating and drinking my whole adult life. When beer was available and flowing freely, I drank it until it was no longer available or until I had drunk enough to put me to sleep. If I saw foods that looked good to me, I ate them. For most of my life, I had not given any thought to the possible consequences that these behaviors could have on my body.

When I was diagnosed in 1995 with Crohn's disease, a chronic digestive disorder which causes the walls of the intestinal tract—the colon, in my case—to erupt in periodic and temporary episodes of inflammation, I was initially terrified that I would have to restrict myself to some bland, formulaic diet, eating pureed vegetables and sipping tepid water. But I quickly learned, from questioning several doctors and from research I did on my own, that—at least according to current medical knowledge—while certain foods may trigger bouts of disease activity in some people, no dietary formula or habits either cause or cure the disease. Modern medicine was unable to pinpoint precisely what caused the eruptions of inflammation in the digestive tract that characterized the disease, but it was relatively confident that food and drink played no part in it. The best current theories held that the disease was the result of an immune system disorder, and that for some reason the immune system was mistakenly directing the body to attack a nonexistent foreign agent in the digestive tract.

For the first few years after my diagnosis, I saw this bit of information as a loophole that allowed me to continue my unhealthy eating and drinking behaviors uninterrupted. Of course I quickly came to realize, from both personal experience and from speaking with a dietitian, that while my diet would not actually *cause* the disease to move into its active state, it certainly could contribute to making my life especially miserable when it already *was* in its active state. In other words, diet could definitely influence the severity of my symptoms when the disease was active—symptoms that consisted primarily of diarrhea, internal bleeding, weight loss, and fatigue. When the lining of the colon is ulcerated and inflamed, the difference on the toilet the next morning between a plain bowl of rice and a bowl of rice with Cajun spices, vegetables, and sausage can be huge.

Hence I was smart enough to watch my diet when the disease was active, but when it was inactive I continued to eat and drink as I had since college—poorly and excessively.

New Orleans represents for me now the last moments of that sort of carefree eating and drinking, and that sort of inattention to the special physical needs of a chronically diseased body. More significantly, but clearly related, New Orleans also represents the last moments in which I really saw myself as distinct from my disease. Laughing in the transvestite bar, strolling along the banks of the Mississippi, eating at the lunch counters of Cajun restaurants—those were the closing moments in the first stage of my understanding of what it means to have a chronic illness, a stage that had lasted nearly five years. It was a stage in which I saw and understood disease as an external invader, as something that came from outside of me and attacked my body at occasional intervals, only to retreat and leave me alone for longer periods of remission. I was like a rebellious high school kid, and the disease was my overly restrictive parents—I behaved while they were around, but when they went out for the evening I threw wild parties and trashed the house.

IF I DIDN'T PAY ATTENTION to my diet, I did at least try to take my medications consistently. The main weapons against Crohn's disease are pills, which come in two kinds: maintenance medications, taken every day, whether the disease is active or not; and medications taken only when the disease is active, in order to help return it to a state of remission.

At the time of our trip to New Orleans I was taking two different medications. The first was Asacol, a topical anti-inflammatory, which means that it remains intact as a capsule until it reaches my small intestine, where it opens and delivers a dose of medicine to the small intestine and the colon. This was my maintenance medication, one that I will probably take every day for the rest of my life.

The second medicine was an antibiotic called Cipro, which in the wake of the World Trade Center bombings and subsequent terrorist acts became known as a treatment for anthrax poisoning. Doctors don't fully understand why Cipro (as it is commonly known) helps some patients with Crohn's disease, but it very clearly has always helped me, so my doctor prescribes it for me as a first line of attack when I am sick. About a month before we had left for New Orleans, I had experienced a slight flare of disease activity, and so I had begun taking Cipro to help calm it down. The pill had been doing its trick, but of course alcohol dilutes the effect of any antibiotic, so in that respect my indulgence in New Orleans might have helped send me down the road toward the more serious flare of disease activity that followed our vacation.

My symptoms were definitely more active when I returned home than when I had left for New Orleans, and had I been able simply to relax and recuperate for a few days I might not have ended up where I did. But as soon as we returned from New Orleans, we had to begin packing for our upcoming family move from Chicago to Massachusetts, and that activity left little time for relaxation. The transition from vacation to packing for a cross-country move probably put such stress on our systems that it would have been unlikely for me *not* eventually to have seen manifestations of it in my body. While stress, like food and drink, doesn't cause the disease, it can act as a trigger for bouts of disease activity, and it can most definitely worsen the symptoms and effects.

Stress and overindulgence in alcohol certainly nudged me toward the cliff edge I was about to fall over, but something else entirely gave me the final shove.

WE WERE MOVING because I had accepted a position as an assistant professor of English at a small New England college. I had spent the last ten years of my life, from my first years of graduate school through three years of an administrative position at my Ph.D. institution, hoping for just such a teaching position. So I had tendered my resignation to the research university where I had spent the past seven years, we had sold our tiny house in a northern Chicago suburb, and we had packed our lives into boxes.

My wife had visited our new city—Worcester, Massachusetts— several times to scout for a place to live, and we had lucked into the purchase of a spacious four-bedroom house just two miles from the college. The house was twice as large as anything we could have afforded within a hundred miles of Chicago.

Because of the children, we took three days to make the drive, stopping overnight at various points along the way. It was the last week in July of 2000, and the weather was beautiful throughout the trip. As we drove through the hilly terrain of upstate New York and western Massachusetts, we were awed by the constantly unfurling landscapes around us: vista after vista of lushly wooded mountainsides.

For the kids, the trip consisted primarily of naps in the car, sing-along tapes, and swimming in hotel pools—with the occasional fit of whining and temper tantrum thrown in, of course. For Anne and me, the trip spurred a combination of anticipation and anxiety about what our new lives would bring—new jobs, in a part of the country we knew very little about, and far removed, for the first time, from much of our families and most of our friends.

The stress factors took a definite turn upward when we finally arrived. We discovered that the previous owners of our new house had taken their washer and dryer with them, not believing them to be "fixed" appliances. We discovered that the moving company had not been able to fit all of our materials on a single truck, and that the second truck would be arriving sometime in the next three weeks. Finally, we discovered that the purchasers of our old house had been unhappy with several things they had found when they moved in—such as a nest of mice in the walls, which I can only guess we had not seen because our cat had kept them at bay—and were threatening to sue us for damages.

During the week we spent sorting all of these problems out, I made one final discovery—I had only a few Cipro pills left, and the bottle I was holding was the last authorized refill I had. To complicate matters further, I was in the midst of switching from my old university's insurance policy to the new college policy I would be on. Until the first day of school in late August, I remained on my old policy. This meant that in order to obtain a refill I would have to call my primary care physician in Chicago, tell him that I needed this medication, have him call a prescription in to a pharmacy in Massachusetts, and convince the health insurance company to pay for it. This last step was essential, since Cipro pills, at that time, cost around $7 apiece; a month's prescription ran close to $400.

In retrospect, none of those steps seem excessively burdensome to me now; at the time, with everything else we had going on, I couldn't bring myself to do it. I was going to stop taking the Cipro at some point eventually, I reasoned—might as well be now.

So, with the biggest stress factor of all just weeks away—the beginning of my first year as an assistant professor, teaching more classes in one semester than I had taught in the last three years combined—I took the last of my pills during the first week of August 2000.

By the second week of August I was in the bathroom ten to fifteen times a day, bleeding, exhausted, and depressed. I was dropping weight quickly and didn't have much of an appetite. I found it difficult to work on my course preparations or even to help much around the house, because of both physical fatigue and emotional apathy.

It's not difficult to imagine how this affected my wife, who was trying to unpack and settle a houseful of stuff, as well as take care of our two children and find a teaching position for herself. I could see her growing increasingly irritated with me, and increasingly unable to handle all of the responsibility that I was dumping into her lap. At the same time, I felt unable to do anything about it.

The arrival of my parents provided a temporary reprieve. They had recently helped my older brother move to Connecticut, and they drove up to Worcester to lend us a hand. My mother was able to help assume some of the childcare duties, which relieved Anne of some of the work. My parents also generally helped around the house, drawing upon their thirty years of experience as homeowners to help us resolve many of the little issues that arise whenever you move into a new house.

They brought as well, of course, their concerns for my health, which, unlike my wife's, were not leavened by frustration at my inability to uphold my end of the household responsibilities. They

urged me to visit a local doctor, but this—like obtaining medicine—would be a complicated affair. I would have to get approval
from my old health insurance provider to see a doctor for an out-
of-town emergency, and then receive special authorization for any
treatment I would need. Again, this does not seem like an excessively burdensome task to me in retrospect, but at the time I could
not muster the energy to do it.

As the days of mid-August crawled by, bringing me closer to a
first day of classes I had now begun to dread, my physical condition
remained fairly stable, but my emotional state worsened consider-
ably.

Until this particular flare-up, I had never paid much attention
to the effects of disease activity on my intellect and my emotions.
Looking back on the two significant flare-ups that preceded this
one, it is evident to me now that I experienced the same bouts of
depression, apathy, and intellectual fatigue with each one of them,
but I had not yet really put two and two together. Certainly I must
have understood that, when the disease was active, I had lost any
desire to get out of the house, do housework, play with my children, read and write, and enjoy my life. But I attributed this simply
to physical fatigue—it never occurred to me that, tired or not, I
should still *want* to do these things, and that therefore some other
factor was coming into play.

My inability to understand that the disease could extend its
reach into my mind and my emotions no doubt stemmed from the
same perspective that had been responsible for my fluctuating diet.
I saw the disease as an external invader, as a biological entity that
attacked my gut, and that could be fought with medicine aimed at
the gut. The disease was distinct from my *self*, and especially from
that part of the self that I valued most—my mind. What could my
colon, after all, have to do with my emotions, my memories, my
desires, my behavior—my sense of who I was?

With the semester looming and no end in sight to the disease activity I was experiencing, my emotional and intellectual state deteriorated precipitously enough to call into question this understanding of my disease.

I felt tired all the time. I was willing to do anything to avoid spending time with my children and assuming my share of the household chores. I abandoned my writing projects. I had no desire to accompany my family to restaurants, sightseeing, or anywhere else. I wanted to stay in the house, lie on the couch, sleep, or watch television.

I could think about nothing but my disease and myself. I was able to cast my mind forward to the very next thing I needed to do or think about, but couldn't push much farther than that. I felt sorry for myself. I was convinced no one understood exactly what I was suffering, and I resented the good health of those around me. However solicitous they were about my condition, I was certain they thought I was exaggerating my health complaints.

I spent my time alternately praying for God to heal me, and resenting Him for striking me with this affliction in the first place. I rationalized not attending Catholic mass, a weekly family event, by pointing out to my mother that the bathrooms in the church were not easily accessible enough.

At some level I must have understood that this behavior was abnormal, and for a while at least I continued to mouth hopeful platitudes and express optimism about my prognosis. But this talk could not ultimately conceal my behavior—and this talk, especially, didn't help get the laundry done, or get the children to bed, or get the lawn mowed, and so on. My failings in these respects were obvious enough, and were creating a cumulative drag on my wife's patience.

One evening during the week of my parents' stay, it suddenly became apparent to me that Anne had had enough. I was tapping weakly away at the computer in the basement, working on a syllabus, and she was angrily folding laundry at a table behind me. We were exchanging irritable words about my inattention to the children, to her, to my parents, and to the household. I was feebly trying to defend myself, when all of a sudden I caught a glimpse of myself, as if on videotape, and the way I had been behaving for the past few weeks. I saw myself on the couch while she and the children headed out to dinner with my parents; I saw myself in bed while Anne straightened up the bedroom around me; I saw myself slumped in a patio chair outside while she played with the children—an astonishing rush of images of passivity.

At that moment I began crying—not so much for my irresponsible behavior, but for the realization that fully appeared to my consciousness at that moment, with all of its terrible and frightening implications. I tried to choke out an explanation to my wife's back as she embraced me, suddenly and mercifully softened by my tears.

"I think this disease is doing something to my mind," I managed to get out between sobs. "I can't understand why I'm behaving like this. This doesn't seem to be under my control."

I stayed locked in her embrace, and then we were both sobbing like children, both temporarily out of control, perhaps acknowledging for the first time that this disease had penetrated our lives in a way neither of us wanted to admit. It had become—at least in those moments when the disease asserted its control over my body—a third party in our marriage, an uninvited intruder that we somehow had to learn to accommodate.

It was a cathartic moment, and an extremely difficult one for me. That a chronic disease can have an influence on one's mind and emotions, as well as on one's physical body, may seem obvious enough, but it was a fact I had been unable to see, or had chosen to ignore, up to that point. In the basement, in Anne's embrace, I finally glimpsed this bit of illumination.

Though it had been just over a month in calendar time, I felt at that moment as if I were a hundred years from our days in New Orleans.

2

A LIFE OF ILLNESS

Learning About Disease

‿

[Diagnosis: Spring 1996; Childhood]

My illumination in the basement that August afternoon was a long time coming—much longer than it should have been. I can best explain why I blinded myself to this insight for so long by telling the stories of how I learned about disease—my own disease, and my older brother's disease, which I had lived with since my childhood.

My illness first manifested itself in the late spring of 1996, with a three-week spell of high fevers, severe diarrhea, internal bleeding, and extreme fatigue. At that time I was twenty-six years old, a new father, and a graduate student in a Ph.D. program in English literature at Northwestern University. I had completed my doctoral exams a few months earlier and had just begun my dissertation.

For three weeks during May of that year, I lay on the couch in my apartment, waiting for signs of improvement in my condition and never receiving any. Before becoming ill, I had been spending two days a week watching my daughter and three days a week working on my dissertation. By the second week of my illness I was sending my daughter to daycare full-time and no longer working on my dissertation. I slept and went to the bathroom. In the morning I woke up with fevers that regularly peaked at 104 degrees, and that would never—despite a steady diet of aspirin—disappear completely.

I tried to beat the diarrhea by not eating or drinking, but it didn't help. One day I can recollect eating a single apple in the morning, and still having to make a dozen trips to the bathroom before going to bed that night. Of course the combination of the fevers, the diarrhea, and avoiding food and drink left me dehydrated, and eventually I wound up in the emergency room with such severe dehydration that they needed to pump several bags of IV fluid and electrolytes into me. My wife and daughter—our first, Katie, who was barely six months old at the time—drove me to the emergency room that night, but left when we realized that I would be spending several hours there. I took a cab home from the hospital at 4:00 AM. As I crawled into bed beside my wife, properly hydrated for the first time in weeks, I felt the first stirrings of hope I had experienced since the illness had begun. Perhaps that had been all I needed to turn myself around.

Within a few hours of waking the next morning, I was back on the toilet.

The doctors in the emergency room, like the physician I had seen after the first week, were baffled by my symptoms. I remember the emergency room doctor sitting at my bedside with a clipboard, posing question after question:

"Have you traveled out of the country lately? Eaten anything unusual? Had shellfish recently? Does your family have a history of problems like these?"

When none of his questions solicited the answers he was expecting, he simply offered me more powerful antidiarrheal medications. Those medications didn't help.

Three weeks into this illness I was finally referred to a gastroenterologist. He saw me, scheduled a colonoscopy, and at last was able to put a name to my condition.

But his diagnosis seemed impossible to me. At the most recent physical I had undergone, my doctor had run through a list of questions about my family history, my habits, and my own medical history. Setting down his clipboard, and rolling up my sleeve to check my blood pressure, he seemed almost disappointed in me.

"Boringly healthy, huh?" he said.

What could possibly explain the transition from boringly healthy to chronic disease in the space of less than a year?

Before the gastroenterologist performed the colonoscopy that eventually helped him make his diagnosis, he outlined for me the two most likely diagnoses for my condition. The first was a bacterial or viral infection of some sort, which would eventually clear up with medication or on its own; the second was Crohn's disease. The causes of Crohn's disease, he explained, were unknown. There was no cure. When the disease was active, medications could be used to help control the symptoms.

I have only one memory from my colonoscopy, which took place under a form of sedation that allowed me to remain semiconscious: at one point I asked the doctor whether he thought it was an infection or Crohn's disease. Smiling, he reassured me that it seemed to him more like the pattern of an infection.

Afterward he gave me my first prescription for Cipro, which cleared up my symptoms within a day or two, and I hoped and expected that I would soon put the entire episode behind me. Within a few days, though, the doctor called me with the results of the biopsies they had performed on tissues taken from my colon, and the news was different. It now seemed to him like Crohn's disease.

I was devastated, especially when I learned that I would now need to take medication daily, whether the disease was active or not.

My doctor softened the blow of the diagnosis by explaining to me that, at least from what he had seen during the procedure, it seemed to him like a relatively mild case of the disease, one that would probably not require hospitalization or surgery (as many patients of Crohn's disease, in the range of half to three-quarters, do at some point in their lives).

In retrospect, I see that my initial inability to come to terms with my disease, and accept its presence in my life, was seriously exacerbated by two moments from this history: the doctor's initial speculation, during the colonoscopy, that I did not have Crohn's disease; and his estimation, once he reversed that diagnosis, that my condition would remain a mild one.

For months after the diagnosis, I held onto the hope that he would again change his mind, that further studies of the biopsies, or my continued good health, would somehow persuade him that I was healthier than he had initially suspected. For at least the first year after the diagnosis, a part of me remained convinced that I could not really have the disease. I did nothing to accommodate the disease, at any rate: I made no changes in my diet; I regularly forgot or skipped my medications; I called the doctor only when I needed prescriptions refilled. I did all I could not to think about it.

Over the next several years, as frequent bouts of diarrhea and bleeding made it increasingly apparent to me that I did indeed have the disease, I held fiercely to the doctor's remark that at least I had a mild case of it. If my case were not serious, perhaps it might find its way into permanent remission? At the very least, I could hope not to suffer from the severe symptoms I had experienced during my first bout.

The second major flare of disease activity, which took place in the late winter and early spring of 1998, helped strip me of the illusion that I had a mild case.

I have two photographs of myself during the nadir of that flare. In the first picture I am in the swimming pool near my parents' condominium in Florida, throwing Katie—then three years old— into the air. She is suspended above me, in mid-toss, and we are both smiling and laughing at the pure joy of it. In the second photo I am at the beach, again with Katie, crouching in the sand and watching a young boy pet a turtle he has found. I am looking up at the camera, a serious expression on my face, squinting against the sun.

What unites these pictures for me is the shock of my naked upper frame. At the time those photos were taken, a two-month struggle with symptoms had dropped my weight from my typical 175 lbs. to around 140 lbs. In the pictures I am shirtless, wearing only my bathing suit, and the effects of the disease on my body could not be clearer.

My ribs, my collarbone, the scapula of my back are clearly visible through my skin. My face is thin and angular, my eyes are sunken, and my nose stands out against the flesh drawn tightly around my face. My head seems precariously perched on a body that hardly seems capable of supporting it. The torso in both pictures is impossibly thin and fragile, as if the slightest blow to my abdomen might crack me in half. My arms are like spaghetti—if I could step into the photograph, I am certain I could complete a circle with my finger and thumb around the upper part of my arms.

I remember standing in the kitchen one morning during that

vacation, trying to soothe my mother's anxieties about my health, to convince her that the situation was not as bad as it seemed. "Look at yourself, Jimmy!" I can remember her shouting at me, tears coming to her eyes as she jabbed at my frame with the spatula she was holding. "You can't see what you look like! You look like you've just come from a concentration camp!" When I didn't know how to respond she turned away from me and wept.

But I could not see myself at that time—could not or would not. Still desperately unwilling to acknowledge the real presence of the disease in my life, I blocked out or downplayed the significance of the weight loss, and the diarrhea, and the bleeding, and the exhaustion I was feeling. All through that flare, and the one that followed it a year or so later, I was able to remain in denial about the disease and its effects on my life.

Looking at those vacation pictures now, I realize that it has only been in the last year or so that I have truly been able to see myself as my mother must have seen me that morning, and as my family must have seen my paper-thin frame angling around the pool and the beach during that vacation—as a chronically diseased person, one whose body, mind, and life course had been fundamentally altered by illness.

The protracted and complicated process of receiving my diagnosis helped close my eyes to this reality for almost five years. But the narrative of my diagnosis does not tell the whole story of my stubborn refusal to accept myself as chronically diseased.

At the age of twenty-six, one comes upon disease with some preconceived notions about what it means, and how to confront it, and how to live with it. For me, those notions came especially from my experiences observing and interacting with the chronic disease that had been in our family for almost as long as my memory reaches back into my childhood: the juvenile diabetes of my older brother Tony, diagnosed when I was six years old.

I WAS IN SECOND GRADE at the time; Tony was in third. I don't remember what symptoms led to the diagnosis, and I don't recollect the experience of the diagnosis as especially traumatic—though undoubtedly my parents, and certainly Tony himself, would have a different perspective.

What I remember instead was how quickly Tony's disease became folded into the routines of our family life. For as long as I can remember, the laundry room in our home served as my father's "doctor's office," where he would take care of whatever small injuries had befallen his children—cuts, scrapes, and splinters, mostly. He had a basketful of first-aid materials he kept on a shelf above the washer and dryer, and he would perch us on the edge of the dryer, carefully examining and tending to our injuries.

When Tony came home with diabetes, the laundry room became the diabetic care center. An extra shelf went up over the laundry sink, one that was quickly filled with syringes, swabs, and small machines—first mechanical, and then electronic—for testing his insulin levels and his blood sugar. Sometimes I would watch Tony administering his treatments; I can remember him sitting on the dryer in the laundry room and testing his blood, giving himself shots—one day in the thigh, the next in the stomach, the next in the buttocks. I remember the pinpricks to his finger, to test the insulin levels in his blood: the click of the tiny mechanical needle, and the bright red drop of blood suddenly and magically blooming from his skin.

When I was a child, those rituals defined Tony's diabetes for me. They contributed to the sense I had—a sense that would one day color my understanding of my own disease—that his diabetes was an external invader, one that he could tame with the regular application of his needle; they contributed, more fundamentally, to my sense that the disease was not essentially part of Tony. Tony lived with our family, and shared bunk beds with me; his diabetes lived in the laundry room.

That sense of Tony as separate from his disease stemmed from other sources as well, one of which was Tony himself, and what I heard from him about his disease.

As a part of his treatment program as a child, Tony regularly visited the Cleveland Clinic for daylong programs called "Diabetic Re-Check." These days consisted of medical testing, meetings with doctors and nutritionists, and activities with other diabetic children. On at least one of these occasions, my parents sent me along with Tony for moral support—I was happy to go, since it meant a day off from school.

I don't recall much from that day except for what I realize now in retrospect was a support group meeting with a child psychiatrist or therapist, in which the diabetic children were to share their feelings with each other about having the disease. That meeting astonished me, because of the depth of feeling I heard expressed by the other children in the group. They talked about the hardships the disease had caused in their lives, and their feelings of isolation and depression. One boy actually spoke about contemplating suicide.

This came as an absolute shock to me. Tony had never expressed such feelings to me—and he and I, so near in age, were close friends—and I had never seen evidence of such emotions in him. Tony's diabetes was simply a fact of life, one more aspect of our family. In that meeting I wondered momentarily whether I had been blind to what Tony was experiencing.

Not for long. After the meeting, he joked about the seriousness of the other kids in his group, and together we laughed at the silliness of the suicidal boy. Why didn't he just suck it up and get over it?

I am certain that at least some small part of this talk, on Tony's part, was bravado, and that he undoubtedly shared some of the feelings that I heard expressed in the group that day. I cannot imagine any eight-year-old child who would not be emotionally affected by the news that he will give himself injections every day for the rest of his life. Tony simply chose not to express those feelings to me, and he protected himself from such feelings that day by mocking them when they appeared in others. He may have been in the same early stages of denial about his disease that I would experience twenty years later.

But I am also certain that Tony really did not feel these emotions, at least on a conscious level, as strongly as the other children I heard that day. I know as well that he faced his disease, as he faces it now, with far greater equanimity than any other chronically ill person I have ever known.

I have a vivid recollection of Tony and myself lounging, one steamy summer afternoon, on the makeshift grandstands that bordered one of our childhood baseball fields. Hot seats, hot sun, a childhood summer day in the middle of an endless stretch of them. Sitting with us were a few other kids from our baseball league, none of them close friends with either of us. They all knew of Tony's diabetes—he toted cans of juice and crackers with him to his baseball games, conspicuous marks of his illness, and his coaches all had to know about his disease in case of an emergency.

That day in the grandstand the other kids were eating candy purchased from the drugstore across the street, waiting for the evening baseball games to begin. For some reason that escapes my memory, the subject of Tony's diabetes came up, and then—in a sudden and inexplicable progression—their initial curiosity about what it meant for Tony to live with the disease turned into something unpleasant.

"Hey Tony," one of the boys said, waving a candy bar in his face, "want some candy?"

"Yeah," Tony replied, with a good-natured chuckle. "Very funny."

The other boys immediately joined in, holding their candy bars out to him and teasing him in the same way for several minutes.

I was furious, but helpless—all of the boys were at least a year or two older than me, and several sizes larger. I had fantasies of smashing my fist into their faces, holding their arms behind their backs and letting Tony do the same. Tony, though, seemed perfectly calm. He responded to the teasing with jokes of his own, and did not seem upset in any way. If there was a delayed reaction, it outlasted our long bicycle ride home; he never, as far as I could see, let the incident disturb him. I asked him about it recently and he had no memory of it.

At the time and even today, in retrospect, his tranquility during this confrontation was surprising to me. That tranquility remained his uniform response to the disease throughout our childhood and even to the present day: I have many memories of experiences with Tony and disease, and not a single one of them involves him showing frustration or anger or depression at his fate.

The source of his equanimity was undoubtedly my father, who refused to allow Tony to see his diabetes as a problem. Tony was simply not permitted to see, to discuss, or to think about himself as disabled or limited in any way. He played baseball, he played football, he ran track and field. He received no special treatment from my parents, at least as far as any of his siblings could tell, nor was he forbidden from any activity in which the rest of us participated.

This attitude toward physical infirmity, disease, or injury was a natural part of my father's temperament, and manifested itself in his dealings with all of his children. I remember distinctly the phrase he used, over and over again, in the face of childhood falls, cuts, and scrapes: "Tough it out." Tony toughed it out like the rest of us, despite the massive difference between our minor wounds and his major chronic disease.

What my father told my brother about his disease, and taught him to think and feel about it, helped Tony cope more effectively than any other diabetic—or any other chronically diseased person—that I have ever known. To this day, my father's impression remains in him. When I started writing this book, I had several conversations with Tony about my own disease, and about how my attitudes toward it had been at least partially shaped by my observations of his experiences. In one of those conversations, I referred to us as sharing the conditions of a "chronic disease."

"You know," he said, "I've never really thought of myself as having a chronic disease. But I guess that's true—diabetes is a chronic disease."

That he could have thought of diabetes in any other way, as anything *but* a chronic disease—one that requires daily injections and monitoring of his blood sugar, attention to his diet, and regular contacts with a physician—is a testimony to the strength of my father's influence.

In an e-mail conversation he and I conducted while I was working on this book, Tony emphasized that learning to think about his diabetes in the way he did was tremendously helpful to him. Though we had not yet discussed our shared memories of these events, he reminded me about the group support sessions at Diabetic Re-Check, and confirmed my memories of what we heard there, and his reaction to it: "You might remember some of the sessions we attended at the clinic where all these kids who could not accept their disease would sit around and talk about it," he wrote. "I remember thinking then that Dad would not stand for this, that they should just 'toughen up.' Perhaps that was insensitive on my part, but it is important in that I did not feel this to be a handicap . . . I cannot tell you how grateful I am for this, especially when I read about the difficulties that others have with their disease."

TWENTY YEARS LATER, in the face of my own chronic disease, I tried initially to "tough it out"—to apply the same principles that had worked so effectively for Tony. But I made a crucial error: My father had taught Tony to face his disease courageously, to forbid the disease from imposing unacceptable limitations on his life, and to accept it as one more of life's challenges. Instead of embracing those noble attitudes, I "toughed it out" by doing my best to ignore the disease, to pretend it didn't exist.

This was no part of Tony's or my father's strategy for handling his diabetes. Tony took his injections religiously, he tested his insulin levels when necessary, and he knew how to regulate his diet to maintain the proper blood sugar levels. Everywhere Tony went that involved physical activity—on the golf course, at his baseball games, on our bicycle rides around town—he carried with him those ubiquitous marks of his disease: a can of apple juice and a small snack. He monitored his condition, and he kept himself prepared for any contingency.

My version of toughing it out led me to behaviors in stark contrast with Tony's careful management of his disease. I ignored or downplayed the significance of early warning signs of an oncoming flare, I postponed telephone calls to the doctor and office visits when I was clearly in need of a medical intervention, and I would decrease the dosages of my medications without consulting the doctor. This was not "toughing it out," as my father defined it and as Tony lived it; this was my inability to accept the presence of the disease in my life; this was denial, plain and simple.

To live as my brother learned to live does not mean foolishly denying that disease imposes limitations on your life; growing accustomed to the necessity of daily injections, and all of the mental baggage that necessity carries with it, certainly constitutes a sort of limitation. To live as my brother learned to live with his disease, to the contrary, means not allowing those limitations to keep us from dreaming about, from striving for, and from doing what matters most to us.

To take medicines every day, to make regular contact with a physician, to submit myself to the occasional colonoscopy, to confine myself to the house or the office occasionally when the disease is at its worst—these limitations have little impact on the things that matter most to me: thinking, writing, reading, teaching, and spending time with my family.

Certainly those limitations are inconvenient, and accepting such inconveniences in my life does not come easily. I hate having to take my medicines, to spill out my pills in the morning or carry them around in my pocket when we are going out to dinner—to pull them out and choke them down, covered in pocket lint, with my dinner water. I don't particularly like visits to the doctor, and no one could possibly enjoy a colonoscopy. I don't expect any of those inconveniences to become less inconvenient with time or with my increasing acceptance of the disease; I can't imagine I will enjoy taking medicine at seventy-three any more than I do at thirty-three. But I have accustomed myself to them, and I have learned to recognize the important role they play in helping me to manage my disease.

What sent me into tears in the basement that August afternoon was the fear that the disease was encroaching upon what mattered to me most—it was preventing me from reading and writing, from caring about and attending to my family, and I had begun to fear that it would interfere with my teaching and research. And indeed, at least for a brief period, the disease had been making inroads into these areas of my life.

I reacted so forcefully because of the sharpness of the intellectual turn I had to take at that moment: from five years of denying that the disease could impose *any* limitations on my life, to the sudden awareness that it was threatening the most essential parts of myself. I was sent spinning from one extreme attitude toward disease to the other extreme.

As I LOOK BACK AT MYSELF in that embrace in the basement, I can see very clearly what I had yet to learn: the arduous and never-ending skill of forging a life between those two extremes. Living on the extremes is the easy part: both total denial and total surrender come easily enough, for they are both strategies for keeping reality at bay. From my diagnosis through New Orleans I lived in denial; for those few weeks in August I lived in surrender.

So, having tasted both of those extremes, I was ready to begin learning the lessons that a year of illness had prepared for me. Those lessons began just a few days after my basement epiphany, where any learning about chronic illness must begin: in the doctor's office.

3

LIVING WITH
UNCERTAINTY

Learning About Doctors
and Medicine

[August–September 2000]

My exchange with Anne in the basement helped me understand more clearly the deeper effects of the disease on my body and my identity, but it also had two more immediate and practical consequences: it opened up a better understanding and dialogue between Anne and myself, and it led to my decision to find and see a doctor immediately, despite whatever insurance hurdles I would have to leap.

Initially, the decision to see a physician gave me a great sense of relief. Once I had decided to take action, I felt confident that, whatever it might end up costing me financially, the proper medications would be able to settle the symptoms down quickly.

31

It was obvious enough to me that the solution to my problem was prednisone, a corticosteroid that doctors use for all sorts of inflammatory conditions and autoimmune diseases, including asthma, arthritis, and Crohn's disease. Prednisone, however, has serious negative side effects, both short-term cosmetic ones (like acne) and more serious long-term medical ones (like osteoporosis). It can also produce steroid dependence, leaving its users unable to get off the drug without sparking a flare-up of illness. Hence doctors generally prescribe prednisone for Crohn's disease in high doses initially to settle down a flare, and then slowly try to taper down the daily dose of the medication until the drug can be withdrawn.

I took prednisone for a year during my last major flare, most of which time was spent trying to taper off the drug, and I had been off it for a year and a half when the disease acted up in August of 2000. I knew that prednisone was what I needed, and I was confident that any gastroenterologist would agree with me. I imagined that my doctor's visit would proceed simply enough: see the doctor, explain my condition and my history, get a prednisone prescription, and wait for the flare to settle.

It's still not clear to me today whether I was being consummately naive, or whether those were reasonable expectations that were not met by the doctors I saw.

I selected a doctor by identifying one from the list of physicians available to participants in my new insurance plan at the college. That way, I reasoned, even if I had to pay for one visit, I could continue to see the same doctor I had been seeing without interruption when my new insurance kicked in. I called the office of a gastroenterologist at a nearby university hospital, and a nurse there scheduled me for an appointment very quickly when I explained the seriousness of my condition. Her reaction was a clear enough message to me that, once again, I had waited too long to take action.

When I arrived, I noticed something curious. The doctor who introduced himself to me, a man who seemed about my age, did not have the same name as the doctor with whom the nurse had booked my appointment. That first appointment proceeded in a strange sort of way. The doctor asked me a series of questions and gave me a brief physical examination. He then excused himself, and was gone for nearly half an hour.

When he returned, I noticed that he expressed his opinions in the form of the royal "we": "We feel that you need to do the following. . . . We would like you to come back tomorrow. . . ."

I realized at that point that I was dealing with a resident, and that in his absence from me he had been discussing my case with some senior doctor, and perhaps with a group of fellow residents as well. This didn't bother me much; doctors need to learn their trade, after all, and I don't mind having them learn on me—as long as they are carefully supervised by more experienced physicians. What bothered me was what he said next:

"We'd like to schedule you for a colonoscopy tomorrow. We want to take a look inside and be certain about what's happening."

"Why?" I said, trying to hide my frustration. "Isn't it obvious what's happening?"

"Most likely," he said. "But it's always possible that your symptoms could be caused by some sort of bacterial infection. We have to rule that possibility out before we can put you on the prednisone."

"Isn't there any other way for you to rule that out?"

The more I thought about it, the less I wanted to schedule a colonoscopy. School was starting in two weeks, and I was not yet fully prepared for classes. My parents were in town, my wife was still hunting for a permanent teaching position, we were readying our older daughter for kindergarten, and I didn't feel like I had time to prepare for, undergo, and recover from a colonoscopy.

The resident responded to my question simply enough: they needed the colonoscopy to rule out the possibility of an infection.

I began to babble out objections.

"I have school starting pretty soon, and my parents are in town, and I have two small children . . . this is *really* not a good time for me to do this."

"I know it's not pleasant," he said, smiling patiently, "but we really have to do this if you want us to treat you."

In other words, we're not treating you without it. No scope—no prednisone.

There was nothing I could do but agree. I thought briefly of trying to make an appointment with a different gastroenterologist, but I didn't have the energy to do that. I was in the pipeline with these doctors, and figured I might as well ride it out.

What bothered me most was the fact that my own knowledge of my body and the disease counted for nothing in this transaction. It was completely and absolutely clear to me that I was having a flare of disease activity, and that I needed prednisone to get myself back under control. In my discussion with the doctor, I had presented that self-diagnosis pretty confidently, expecting to find agreement and a prescription.

I should make clear that I do not normally diagnose my own medical conditions; if I suddenly began to have severe headaches, I would be in the doctor's office immediately, looking for a diagnosis. But living with a chronic disease is different. I live with the disease on a daily basis; nobody knows its symptoms, its patterns, its different manifestations in my body as well as I do. I will run to the doctor as quickly as anyone when something new appears, but

I had been through two flares before this one, and was experiencing exactly the same set of symptoms that I had seen before.

Subsequent events turned out to support my interpretation—I was, of course, having a flare of the disease. But the doctors did not count my interpretation of what was happening to me as medical knowledge; instead of listening to me, and perhaps calling my previous physician to confirm my history, they elected to perform a highly invasive and unpleasant medical procedure to test their own interpretations of my condition.

And perform it they did—my resident friend attending, a more senior doctor actually conducting the procedure. They almost immediately decided, once they had begun, that my colon was so inflamed that they were unable to proceed very far for fear that the scoping instrument might puncture the colon wall. So the procedure lasted less than a half hour, though that brevity did not in any way reduce the unpleasantness of the preparation and the recovery, both of which rival the procedure itself for their ability to induce physical discomfort.

Sometime during the hour or two I spent recovering on a cot in the hospital, someone handed me a prescription for prednisone. My dad and I filled it on the way home from the hospital and I began taking the medication immediately.

A week later, I had a follow-up visit scheduled to discuss the results of the colonoscopy. Once again, I was seeing the resident.

"Well, the tests for infection were all negative," he said, consulting my folder. "So you can go ahead and begin taking the prednisone."

"I've been taking prednisone since the day of the colonoscopy," I said.

"Oh," he responded. He cocked his head and gave me a suspicious look, as if he were evaluating my trustworthiness as a patient.

I could only assume that either nobody had told me not to begin the prednisone yet, or that they had told me while I was still groggy from the procedure, and it hadn't registered. But he didn't say anything else about it. The rest of the time we discussed my medication schedule and set up one more follow-up visit.

As I drove home from that appointment—on a beautiful August day, one of those late summer days that make it seem so *unreasonable* to be sick—it again occurred to me that Jim Lang had been an insignificant part of this transaction. At times it seemed almost as if the dialogue about my health were taking place between the doctor and my colon. The doctors working with me were fully aware of the seriousness of my symptoms, and of my life situation—the resident had shown a mild interest in the fact that I was a professor, and had asked me a question or two about my upcoming academic appointment. He knew that I was preparing for my first year as a professor, that we had recently moved, and that I had small children.

Nonetheless, he and his colleagues had been willing to let a week pass without offering me any treatment whatsoever—a week of a dozen bowel movements a day, debilitating fatigue, and continued weight loss. Somehow that part of the disease, the effects that it actually had on the patient's life, seemed to fall outside their purview.

Doctors are not therapists, I know. It was not my resident's job to hold my hand, help me prepare my syllabi, and talk me through my emotions. But the lack of understanding and sympathy I experienced at the hands of this team of physicians was an eye-opener for me. For them, I was a diseased colon—nothing more, nothing less.

At my final follow-up visit, to which—thanks to a babysitting mix-up—I had to bring my then-two-year-old daughter, Madeleine, I had one last disturbing exchange with the resident.

Reviewing again the results of my colonoscopy, he explained that the group of doctors who had been consulting on my case were all in agreement that my illness seemed to them more likely to be ulcerative colitis than Crohn's disease.

I knew from my own research on these related diseases that ulcerative colitis differs from Crohn's disease in that it is limited exclusively to the colon, and can be cured by a colectomy—complete removal of the colon. This doesn't work for Crohn's disease; when the colon is removed in cases of Crohn's disease, the inflammation may simply spread up the remainder of the digestive tract. Ulcerative colitis also primarily affects the inner lining of the colon, while Crohn's disease affects the entire wall of the inflamed portion of digestive tract. Apart from these subtle distinctions, though, Crohn's disease in the colon—or Crohn's colitis—and ulcerative colitis do bear a strong resemblance to one another.

The resident explained that a variety of factors suggested to them the alternate diagnosis of ulcerative colitis. I had heard this possibility before, from two previous gastroenterologists, but in the end both of those doctors had come down on the side of Crohn's disease.

I pointed this out to the resident, and noted the reasons my previous gastroenterologists had given me to support their diagnosis. The antibiotic Cipro, for example, which had always been an extremely effective drug for me, was known to work primarily for patients with Crohn's colitis. How would his diagnosis explain that?

"Well, we're not really sure," he said. "That would definitely suggest more of a likelihood of Crohn's."

"So how can you tell?" I replied. "And what difference does it make?"

"We may not ever be able to tell for certain," he said. "And in

fact, around ten percent of cases of colitis can't be definitively classified, and are referred to simply as 'indeterminate colitis.' You may
be one of those cases."

"So what difference does it make?"

"Not much in terms of your treatment right now," he said,
"but we can ultimately cure ulcerative colitis."

"Yeah," I said, a little taken aback at that remark, "but I'd prefer to keep my colon."

"Don't worry," he said, with a small laugh. "You're not at that
stage yet."

At that point he excused himself, and I waited in the tiny
examination room for another forty-five minutes. Madeleine had
sat patiently enough through my conversation with the resident—
who had ignored her completely—but this waiting period pushed
both of us to our limits. By the time a half hour had passed, I was
letting her play drums with the tongue depressors, wheel madly
around the room on the physician's stool, and pull yards of sanitary
paper across the examination couch.

Finally a different doctor came in. He was an older man, no
doubt the supervisor of the resident(s) who had been studying my
case, but he made no mention of this to me, or of his role in my
treatment. No one ever explained to me that I was being treated
by residents, or told me who was the supervising doctor, or anything at all about the structure of care I was receiving. This new
doctor had not been the physician who performed the colonoscopy; I had not seen him before in any context.

His role was apparently to put closure on my case. Because of
a change in my insurance plan, it turned out that I would not continue seeing these physicians, so I had informed them that this was
my last visit. This senior doctor spoke with me for several minutes
about my medications, and about the future course of my treatment, and then sent me on my way.

These exchanges during my final visit with the doctors disturbed me more than any of my previous visits—though for different reasons. After five years with the disease, I had grown
accustomed to doctors and their different ways of treating
patients—I experienced nothing with those university physicians
that I had not seen before, if perhaps in less intense forms.

What I found much more intolerable was the lack of certainty
about my condition that I was forced to confront, and the realization that that uncertainty may never be resolved. It was not specifically the inability of my doctors to put a name on my condition
that bothered me; I can hardly fault them for their unwillingness
to make a definitive diagnosis in the face of a case that doesn't
follow the rules. And I'm not sure, ultimately, if it makes a huge
difference whether I have Crohn's disease or ulcerative colitis, since
the treatment for both—up to the possibility of surgery—is essentially the same.

It was, instead, the way that this inability to diagnose my disease conclusively changed my understanding of the practice of
medicine.

I suspect that most people come to their first conscious understanding of doctors, and medicine, at some point beyond childhood, as I did: Doctors have a definite knowledge of the human
body and its potential defects and diseases, and we have only to
consult the correct doctor to find the solution to our bodily mal-
functionings. Certainly the body can have problems that exceed the
curative powers of modern medicine, but even in those cases, such
as cancer, doctors *know*: They can make clear diagnoses of our
problems, they can see and understand which problems can be
resolved and which cannot, and they provide the proper resolutions
when they can.

One hears occasionally that many doctors have a "God complex," believing they can heal the sick and raise the dead; I suspect that this perception is more prevalent among patients, especially those who have not had much experience with doctors and disease. Our society assigns doctors especially revered places of authority and respect, and that reverential treatment depends largely upon our faith in their definitive knowledge of the human body and its ills.

The surest way to cure a patient of these perceptions is to put him in close and frequent contact with doctors trying to deal with a case of chronic illness.

My exchange with the resident on that final visit reminded me of another exchange I had—baffling to me at the time—with the physician who initially diagnosed me with Crohn's disease.

Or . . . sort of diagnosed me with Crohn's disease.

Several days after the colonoscopy he had performed on me, he telephoned me to report the results of the biopsies they had taken from my colon. Originally, just after the procedure, he had told me that it had seemed to him more like a bacterial infection than Crohn's disease. On the telephone, he explained that the tests were suggesting a different diagnosis.

"Unfortunately, Jim," he said, "what I saw in the biopsies was more consistent with Crohn's disease than with a bacterial infection."

"Okay, so . . . what exactly does that mean?"

"It means that your symptoms and tests are all consistent with what we know about Crohn's disease."

"So does that mean I *have* Crohn's disease?"

"The symptoms and tests are all consistent with Crohn's."

When I finally hung up the phone, it still wasn't clear to me whether I actually had the disease or not. I repeated the conversation to my wife, and we agreed that his strange word choice, evasive and ambiguous as it sounded to me, did seem to suggest that I had the disease. But I couldn't understand why he had not simply come out and said this: "You have Crohn's disease." Was this some strange doctor-speak, a way of avoiding medical malpractice suits in case his diagnosis turned out to be mistaken?

Looking back on that conversation from my final office visit at the university hospital, it was much clearer to me what he was doing: He was unwilling to commit definitively to a diagnosis for the simple reason that he couldn't be certain of his diagnosis. This had nothing to do with his personal competence as a physician; it had everything to do, instead, with the nature and limitations of medical knowledge in the face of this chronic disease.

Previously the relationship between the human body and its physicians had seemed to me parallel to that of an automobile and its mechanics. The parts of an automobile could malfunction in all sorts of ways, but those ways would never exceed the knowledge of the best mechanic. Whether or not he will be able to fix the problem, the mechanic will be able to identify and understand it.

What separates the automobile from the human body, of course, is that human beings design and construct automobiles; human beings reproduce other human beings, by contrast, but they do not design and construct them. The auto mechanic can always consult the manufacturer's design of an automobile to help him make his diagnosis; physicians have no such document to help them make their diagnoses.

We expect physicians to have answers. For most of my life, whenever I was ill, I viewed a trip to the doctor with relief. After seeing the doctor, I would have a clear diagnosis and probably a prescription as well. Even those people who avoid and fear seeing doctors probably do so for a similar reason—the doctor will give them a definitive explanation of their ills. The reluctant patient simply doesn't want that definitive knowledge, out of fear that the news will be bad.

Learning to live with Crohn's disease, by contrast, has meant learning to live without definitive answers, and with uncertainty. This uncertainty reaches from the very nature of my condition itself—bacterial infection or Crohn's disease? Crohn's disease or ulcerative colitis?—to the nature of the different treatments and medications that have been prescribed for me. The medicines I take at the moment are different from the medicines I took six months ago, and probably differ from the medicines I will be taking six months from now. My doctors and I are constantly tinkering with the kinds and dosages of the different medicines that are available for Crohn's disease, searching for the combination that will keep me in sustained remission. It sometimes seems to me that I am the subject of an experiment designed to determine how many different medications the human body can withstand over the course of a year.

Learning to live with Crohn's disease has meant learning to pay close attention to the physicians I see, and learning to think for myself about which tests, treatments, and medications make sense to me and which don't. It has meant conducting my own research on the disease, and keeping up to date with new developments and treatment options. It has meant learning to think for myself about which doctors can provide the different kinds of help I need.

Several years ago I was seeing a new primary care physician for the first time, and listing the medications I was taking at that time. When I mentioned one of those medications, he stopped me and asked me a few questions about it. Then he walked over to his desk and opened his *Physician's Desk Reference*, apparently consulting it for information about this particular drug. I realized at that moment that, if I needed to change my current medications or add new ones, this wasn't the doctor I wanted to consult.

But the need for me to think clearly about my disease, pursue information outside of the doctor's office, and make decisions on my own became most clear in the indirect dialogue I witnessed between my gastroenterologists and the nutritionists with whom I have spoken over the past several years.

I HAVE WORKED OR SPOKEN about my condition with five different gastroenterologists over the last six years, and I have asked each one of them the same question: "Should I be restricting my diet in any way? Are there any foods or drinks I can eliminate or add to my diet that will help my condition?"

The answer has been completely uniform: Diet has nothing to do with this disease. Eat and drink whatever you like.

This, of course, was the answer I wanted and liked to hear. Having no restrictions on my diet lessened the impact of the disease on my life.

However, in the course of conducting my own research about the disease, I have come across a number of books, articles, and people that have given me the opposite answer. Some of these sources have been diet books; some have been academic studies that suggest, for example, that consuming certain kinds of fish can help reduce inflammation in the body; and some have been nutritionists and dietitians whom I have seen, at the urging of my wife and family, over the past six years.

Anne has consistently encouraged me to see nutritionists about my eating and drinking habits, on the theory that we have to explore all possible means of treatment, and I have always been willing to do so. I view any conversation with a medical specialist as having the potential to add to my store of knowledge about Crohn's disease, about the body in general, and about the practice of medicine.

While no nutritionist or diet book has ever claimed that any one particular diet can cure my disease, or put me into permanent remission, they do consistently claim that diet can influence the course of the disease: It can help to induce remission, it can help to maintain longer remissions, and it can help to relieve symptoms when the disease is active. These claims are seductive, because they give me a sense of control: If I follow these dietary prescriptions, I can actually reduce the severity of the disease. If I listen to my gastroenterologists, by contrast, I have no such control—I remain in a purely reactionary mode, responding with medications whenever the disease becomes active.

How on earth does a layperson adjudicate these competing claims about treating the disease? I expect my gastroenterologist to have a fuller and more complete knowledge of the disease than a nutritionist; at the same time, I expect a nutritionist to have a fuller and more complete knowledge than my gastroenterologist of how diet and nutrition, at least on a general level, contribute to one's overall health. The gastroenterologists' perspective means I can eat and drink as I like; the nutritionists' perspective gives me some power to help keep my disease under control.

The dilemma of having to reconcile or choose between these competing perspectives reflects precisely what I have learned this year about doctors, about medicine, and about living with uncertainty. I have come to two important conclusions from my experiences dealing with physicians and my chronic illness: one philosophical, and one practical.

Philosophically, I have to accept that my physicians do not have the answers to all of the questions about my disease—more disturbing still, that those answers simply do not exist. The human body does not present itself to my physicians—to any of us—as an open textbook. It is, instead, like everything we find in the natural world: complex, mysterious, and capable of bursting through the categories of knowledge we have constructed to contain it. It does not obey laws or rules, it does not always respond to medications in the ways we expect it to, and it evolves and changes continuously.

Practically, this means that no one else can make my medical decisions for me. The input of friends, family, and medical specialists all can help to inform my decisions, but ultimately I have to take responsibility for them.

Consequently, all of us faced with chronic diseases need to become medical researchers, gathering as much information as possible about our bodies, our diseases, and the different treatment options currently available to us. In every interaction I have with a health care professional, I ask questions—not searching so much for answers as for perspectives. I want to understand all of the different possible ways that my nurses and physicians might be thinking about my body and my disease, so that I can identify specific treatment strategies when they are presented to me and have already thought through them. I do research on the medications prescribed for me, and read up on the latest developments in the research and treatment of my illness.

Those of us with chronic diseases need to let our knowledge inform our interactions with our doctors. This does not mean rushing into the doctor's office with the most recent newspaper articles on miracle cures. It also does not mean avoiding the doctor, and instead trying cures we read about on the Internet or hear about from friends. It means, instead, learning to make informed decisions about the care one is receiving, from choosing the right doctors to helping decide upon the best possible courses of treatment.

IN THE DIALOGUE BETWEEN my doctors and the nutritionists over the role of food and drink in affecting the disease, I reconciled their competing perspectives by acknowledging that diet certainly can help control and relieve my symptoms—if not the disease itself—and for that reason alone, I should pay some attention to it. But I also decided that I did not want to sacrifice completely the foods and drinks I loved.

So I have compromised. I place only a few food and drink restrictions on my diet, but I do make sure to eat certain healthful foods regularly, and to take vitamin supplements to counterbalance the effects of some of my medication (such as calcium pills when I am taking prednisone). This way I am ensuring that I receive the benefits of these foods and vitamins, while not forbidding myself the foods and drinks I have always enjoyed. I have also discovered recently that soluble fiber can be extremely helpful in thickening up my bowel movements and controlling diarrhea, even during moderate flares. So I eat oatmeal for breakfast every morning, and have a dose of Metamucil before lunch and dinner.

The right answers? The proper attitude toward diet and nutrition?

I wish I knew.

4

DISEASE'S ROUTINES

Learning About Daily Life

~

[October 2000]

Mornings are the worst.

Not simply because I hate getting up in the morning—which I do, and have done for most of my adult life. I am an evening person, and rarely make it to bed before midnight. I am in the right profession for this nocturnal preference. I can usually schedule my courses to begin in the late morning at the earliest, and I can do much of my schoolwork at home, on my computer in the basement, throughout the evening hours. During the school year I am often at work from 8:00 PM until midnight during the week, after a full day of classes and office hours.

However, at this point in my life, mornings have become a necessary part of my routine. My children are two and four years old, and require transportation to school and daycare; my wife teaches elementary school; and, by some bad scheduling luck, I have an

8:30 AM class two days a week this semester. So I am getting up early in the morning, at least for the fall semester.

And here's the problem: When the disease is active, mornings—and early mornings especially—are the worst time of the day for my bowels. I have to wake up at least one hour before I leave for school in the morning, to give my system a chance to excrete everything it wants to excrete before I can trust myself to the ten-minute car ride to my office at school. That may translate into as many as four or five bowel movements before I leave the house, depending on how I'm doing on any given day.

On this day in particular, a Tuesday in mid-October, I wake up around 6:30 AM. I like to be in my office at least forty-five minutes before class starts to collect my thoughts and have a few moments to gather my energy before I have to face fifteen eighteen-year-olds at 8:30, most of whom are dazed and bewildered, wondering what they're doing up so early. I wonder whether they know I hate it as much as they do.

Anne's already out of the shower by the time I pull myself out of bed at 6:45, and I stumble around her, grab my pill bottles, and spill out my morning medications: four Asacols, one Cipro, and two ten-milligram tablets of prednisone. I slug down the Asacols immediately, with a small cup of water, but the Cipro requires a full glass of water and the prednisone has to be taken with food. I carry them down with me, where I find my two daughters up unusually early, perched on the couch in front of the television. They must have heard Anne waking up and come out to find her; she must have turned on a television show for them.

I say good morning—they ignore me, transfixed by the television—and find my way to the kitchen. By the time I get there, I have to make my first trip to the bathroom. It's diarrhea, of course, but there's no blood. That's a good day. Sometimes what I want more than anything are the simplest things of all: not to have to look at my excrement in the toilet, reading it like some kind of fortune-teller for signs of an oncoming flare. I would love to be able to sit on the toilet, wipe myself, and walk away.

Back in the kitchen I pour a full glass of water, grab a banana—the only raw fruit or vegetable I allow myself to eat right now—and slump down at the kitchen table. I pop the Cipro in my mouth, gulp down as much water as I can, and begin eating the banana. When I'm finished with it, I take the prednisone, and pour myself a glass of orange juice.

I carry the juice upstairs to the bathroom with me, which Anne is just leaving. Another stop at the toilet before my shower—again diarrhea, again no blood. I feel that I might have one more bowel movement in me before I have to leave, and then might be finished for the morning. Feel or hope? Not sure.

Shower, shave, dress, and back downstairs for my morning chore—making lunch for Katie, my four-year-old kindergartner. The girls are seated expectantly at the table, still in their pajamas, and Anne is cutting up strawberries for them and making toast for their breakfast. I pour them glasses of milk and work on Katie's lunch.

Katie would make a good Crohn's patient. She eats the same bland, boring foods day in and day out: peanut butter and jelly sandwiches, macaroni and cheese, crackers, bananas. As Anne serves the girls their breakfasts, I want to reach over and grab one of the leftover strawberries and pop it into my mouth. Though I haven't eaten one in months, I can remember the taste of them—the tiny seeds, rough along your tongue and teeth, the sweet and sugary juice.

It's not worth it. I would pay for it later today, or tomorrow morning.

After the girls are finished eating I run them upstairs to help them get dressed. Two-year-old Madeleine wants to wear only dresses, no matter what the weather, and only certain ones at that, so I throw some clothes at Katie, who can dress herself, and begin the morning's battle with Madeleine. We finally compromise on a dress with a turtleneck underneath, and I help her into her clothes.

All of us back downstairs, the girls settle down for a half hour of playing before they have to leave for school. Anne and I share a few words about our schedules. Tuesdays and Thursdays are complicated childcare days, since I have to be at school from 8:00 AM until at least 4:00 PM. Anne drops both girls off every morning—Katie at school, Madeleine at her in-home daycare. On Tuesdays and Thursdays she picks up Madeleine from daycare after her school lets out, and Katie, whose kindergarten class ends an hour before Anne gets home, walks home with an older neighbor girl, and stays at her home until Anne picks her up.

On Mondays, Wednesdays, and Fridays I have no classes, just office hours from 10:00 AM to 2:00 PM. On those days I pick Katie up when she gets out of school, and together we pick up Madeleine.

I am ready to leave this morning, but stalling near the door. Do I make one more stop in the bathroom, just for safety's sake? I don't feel any urge at the moment, but that could change in a heartbeat—literally. I watch a few leaves fall from a tree in our front yard, hesitating. Up and down our tree-lined street the colors of a New England fall are in their full glory. I think for a moment how nice it would be to take a leisurely drive around the area, admiring the foliage, if only I could be assured of not having to dash off to the side of the road in search of a restroom at any moment.

I take a chance and head out the door. We live 2.3 miles from the college, a blessing for which I could not be more grateful. I barely get to the end of our street before I begin to feel some pressure in my bowels. Turn around or keep going? Can I make it the seven or eight minutes it will take me to get to school?

I press onward.

Almost a mistake. Some traffic at the intersection of the grade school I have to pass slows me up, and when I finally pull into the parking lot the pot is nearly boiling over. I dash into my office, throw my stuff into a chair, and make it to the bathroom just in time. The stalls around me are empty at this time of the morning, and it's a relief to be at school, so I relax and take my time. Colleges are wonderful places for people with Crohn's disease to work—public restrooms everywhere you turn. I'm never more than a few hundred yards from a toilet.

Anne firmly believes that my close calls in the car, which happen frequently, are a psychological phenomenon—that every time I get into a car I suddenly have to go to the bathroom, because I am paranoid about finding myself stuck away from a public restroom. The fact that I almost always—though there have been a few notable exceptions—make it to a restroom on time would support that theory. I don't doubt that Anne is correct, but even if the source of the pressure on my bowels is psychological it doesn't make it any less real, or any less insistent.

Back in my office I pull out the small scrap of paper I have been carrying around with me for the past two weeks and write today's date on it, along with the times of each bowel movement I have had that day. I mark a small notation, "NB"—no blood—next to each of them. The scrap of paper is covered with tiny, cryptic writing: the dates for the past two weeks, under which stretch long columns of times: 7:00 AM, 7:20 AM, 8:00 AM. The columns, at this point, are mostly five to seven entries long. I have had columns that stretch into the high teens.

One of the best weapons against disease, I know, is information. The more information I have about what constitutes normal for my body, the more quickly I will be able to identify deviations from that normalcy, and know when to seek treatment. So I should keep these notes all of the time—instead of, as I do, in clumps of two- or three-week periods, saved on random scraps of paper I shove into my wallet or drop into my desk drawer at work.

Although I know I should keep this notebook constantly, I find I can only tolerate it for so long. At some point, seeing the columns slowly fill the piece of paper begins to depress me, so I toss the note into a file folder I keep on the disease and prefer to live in ignorance for a while.

In the office I settle in at my desk, check my e-mail, and prepare some handouts for class. At 8:15 AM I am ready for one last trip to the restroom when a colleague stops into my office for a chat. She has nothing in particular to say, just idle chatter. Given the solitary nature of the academic life, these interoffice visits are often the only opportunity we have for conversation with other adults during the workday.

But as the chatter stretches on, the pressure on my bowels begins to increase. I hold, clench, breathe slowly, and begin fooling around with papers on my desk, hoping she'll get the message that I need to excuse myself.

If my disease were public knowledge, I could easily interrupt and tell her what was going on. But thus far—two months into my first semester—I have decided to keep my disease a secret. Part of that stems from not wanting others to see or treat me differently; the other part stems from the socially embarrassing nature of the disease's symptoms. Somehow I find it difficult to tolerate the thought of excusing myself from a conversation or a meeting, and knowing that everyone in the room is wondering if I am running off to have a watery bowel movement.

Eventually I tell her that I have to leave for class—this she can certainly understand—and I gather my things and take them with me for one last trip to the restroom.

After four or five bowel movements since I have been awake this morning, I can tell I am getting dehydrated. My mouth is dry, and I feel slightly shaky. I get a bottle of water from the vending machine near my classroom.

In class—freshman composition—the students are workshopping first drafts of an essay, which means they sit in circles of three or four and exchange their papers, reading one another's work and responding to a series of questions that I have designed for them.

In other words, easy day for me. However, I usually have to spend ten or fifteen minutes beforehand going over course business—upcoming due dates, changes to the syllabus, announcements for one or another event sponsored by the department—and preparing the students for the exercise.

About ten minutes into this preparatory patter—no surprise—I begin to feel some pressure on my bowels once again. I speed up my delivery. If I can make it through a few more minutes, I will be able to run to the restroom across the hall while they work on one another's papers. Pressure builds slowly, and I think about making a dash for it before I have even finished, but then it lessens for a few moments. I finish up my comments, see that they have begun reading one another's papers in earnest, and get to the restroom with time to spare.

While I'm there I feel gratitude again that I teach at the college level, in classrooms which one can leave unattended for a few moments without fearing that the students will burst into riots or destroy the room.

Sometimes I run through different professions in my head, wondering which ones would be the most difficult for a person

with Crohn's disease—for a person, more plainly, who regularly
has to make sudden and urgent trips to the restroom.

Best jobs for Crohn's disease: any job in a big office with large,
anonymous restrooms—though it should be a lower-level position
that does not involve spending an excessive amount of time at high-
pressure meetings; any sort of freelance creative work; any job that
one can do at home; health professional (hospitals and doctor's
offices are chock-full of bathrooms); hotel employee (ditto); and,
of course, restroom attendant.

Worst jobs for Crohn's disease: sales (lots of traveling and
meetings with people); sole proprietor of a store; airline pilot; professional athlete; Wall Street trader ("I had to run to the restroom,
and I lost my firm $100,000"); and, perhaps worst of all, surgeon
("Nurse, put your finger right here and press down until I get
back").

In class everything winds down smoothly, and I finish up the
rest of the morning without incident.

Back in my office, I settle into the easy chair I bought at the
Salvation Army and allow myself a few moments of depression. It
has been a bad morning. Close calls on the way to school, in my
office, and in the classroom.

Close calls always intensify my feelings of stress and depression
about the disease, because not all close calls are happily resolved.
The offices in my department are in former dormitory rooms, and
each office has a student wardrobe in it. In that wardrobe I keep
an extra pair of shoes and a complete change of clothes, in case I
don't make it to the toilet on time.

This has not happened to me more than a handful of times,
but each instance of it sends me into a spiral of depression. It is
almost impossible to convey to the uninitiated the full sense of deg-
radation, shame, and humiliation one feels after an accident like
that.

The most recent accident I had occurred with only one witness—my youngest daughter, Madeleine, who never even knew that it had happened.

IT HAPPENED ONE FRIDAY early in the fall semester, when I had no classes and had elected to keep Madeleine home with me instead of sending her to her daycare provider. I had planned a trip to a nearby mall to purchase birthday presents for my wife.

I waited until nearly 10:00 AM to leave, wanting to ensure that my bowels were completely emptied for the morning. When we finally took off I was confident that everything was fine, and the two of us set off in high spirits, singing along to one of her tapes of children's songs.

Not five minutes into the ride—just, of course, as I was pulling onto the highway—I felt a sudden wave of pressure. I thought I could hold on until we made it to the mall, which was a mere fifteen minutes away. I didn't have much choice, at any rate—we were still new to the area, and I didn't know the highway exits well enough to locate a public restroom quickly if we did exit.

Five minutes into the ride, the situation had begun to reach crisis point. No exit signs were in sight, and along the side of the road—the absolute last resort for a desperate Crohn's patient— there was no shoulder and no tree cover. On either side of the traffic lanes were thin shoulders, concrete barriers, and sheer drops down into a valley below. I couldn't just pull over, and even if I could I would have had to take care of my business in full view of all passersby.

I did my best to hold on, but it is literally impossible to hold material in your bowel when it is inflamed and bloodied; the body eventually will force it out, as it did at that moment. Some newspapers were lying on the passenger seat, and I reached for these and put them under myself to preserve the car seat as best I could. Usually releasing only a small amount will relieve the pressure, so I wasn't exactly swimming in the stuff, but it was enough to make me uncomfortable. I pulled off at the first exit, and got back on the highway heading for home.

Twenty very long minutes later, I pulled into my driveway. Madeleine had fallen asleep. I glanced at the neighbor's house and saw no signs of life, so I ran quickly from the car, through the garage, and into the laundry room. I stripped, wiped myself down, and found a dirty pair of sweatpants on the floor. I put them on, retrieved Madeleine from the car and deposited her—still sleeping—on the couch, and took a shower. In the shower I couldn't help but cry a little bit, feeling sorry for myself. It was part shame—Why?—part relief—What if one of the neighbors had seen me? What if it had happened in the middle of a store at the mall?—and part frustration and rage at the hand I'd been dealt.

To sit in your own shit is to come face-to-face with the ugliest part of yourself and your body, and to be given a visceral reminder of the disease's power to control your life. I can ignore my diet, I can refuse my medications, I can even choose to end my life if I want—but when that stuff has to come out, it comes out, whether I want it to or not. No choice; no free will; no control.

This is a jarring lesson to receive on a sunny Friday morning in September on the way to the mall—jarring at any time, I suppose.

The one blessing buried in the experience was that Madeleine never knew that anything unusual had happened. She never even asked me why we didn't make it to the mall.

I SHOULD TAKE A LESSON from Madeleine's obliviousness, but instead—back in my office easy chair on this sunny October morning—I replay the incident over and over again in my mind. The thought of that day still incites feelings of shame and humiliation, and my mind hovers around the memory the way the tongue endlessly probes a canker sore. It is the sort of memory that one tries to shake off with a jerk of the head, but that lingers on, floating in some place in my mind that I can't seem to close off.

On the positive side, I feel a certain amount of gratitude that I made it through this particular morning without such an incident.

That sense of gratitude may be one significant thing that separates the chronically diseased from the rest of the population—we can be grateful for the smallest of life's gifts. You can't truly appreciate a solid bowel movement until you've gone six months without one. You can't truly appreciate a public restroom until you've shat yourself for want of one.

I fill the remaining hour or so until my next class doing one of the few things that makes me feel better about the disease—reading the stories of others who share my problems on one of the many Web sites devoted to Crohn's disease and ulcerative colitis.

I find a good one this morning:

> Traveling to England as a professor for an adult oriented college trip, I suffered intense flare-ups from my newly diagnosed Crohn's Disease. My good friend, the director of this Lutheran University, and I were chaperoning students to Bath, England. During the morning of sightseeing many cathedrals, I felt the intense urge to find a restroom. Vicki, knowing my condition, started looking for a WC (water closets as they are called in Europe). I am now running through the streets of Bath, deserting the students in the cathedral—un-chaperoned. Vicki and I find a pub—I proceed to make the WC my home for the next 1/2 hour. When I finally arrive out of the WC, I find my friend Vicki, having a 1/2 pint at the pub's bar. You see the owner of the pub demanded that only patrons could use the restroom,

and that since I was a rude American running into the bathroom, she had to purchase something from the pub. What
makes this story interesting is that it was 9:00 in the morning
and the students were all standing in the doorway looking for
the two chaperones and finding us having a beer.

This post comes from one of my favorites: the IBD Humor
Page.* Compiled by a sufferer of Crohn's disease, the site contains
the stories of dozens of people who have found themselves in situations as—or more—embarrassing than any I have ever experienced. Knowing that others share the same tribulations brings a
comfort of sorts. The professor's story brings me particular reassurance because I have recently agreed to co-chaperone a group of
students to Ireland for spring break next year.

But in general, the contributors to these pages have learned a
lesson that took a couple of years to sink in for me: It's hard to
take oneself too seriously when you've sat in your own shit as an
adult—and on more than one occasion.

Revitalized and comforted by the stories of others—and
another dose of Asacol working its way through my system—I have
a fantastic class in contemporary British fiction, in which we are
reading and discussing Grahame Greene's novel *The End of the
Affair*. Near the end of the novel the main female character takes ill
in one of those unspecified, novelistic ways. She has a cough, and
becomes weak and feverish, and then dies in her bed. How glibly a
novelist can gloss over the painful details of illness! It makes me
wonder whether Greene ever came into close contact with real illness—if he had, I'm not sure he would have been so willing to use
the character's illness, killing her off with a cough and a fever, to
provoke some self-reflection on the part of his male narrator.

*This site was no longer active at the time this book went to press.

I found this tendency on the part of writers about illness especially frustrating during the few years between the time I was diagnosed and the time I discovered Web pages like the IBD Humor
Page. I would read about cases of Crohn's disease patients in published books who would describe their troubles with easy euphemisms: "I was having a lot of GI trouble at that time. . . . My
stomach was really acting up. . . . I was having a lot of discomfort
in my abdomen. . . ."; or, the blandest of all: "I was sick."
    What do you mean by that? I wanted to shout to the books I
was reading. Have you soiled your pants like I did? Have you spent
twenty awful minutes on the toilet, and then stood up and looked
into the bowl and seen your own bright red blood? Have you lain
in bed late at night, knowing that inside you somewhere, in a place
you can't reach and can't control, blood is leaking from your intestines? Have you lain in bed late at night and wondered whether the
disease is slowly burning holes into the wall of your intestines as it
sometimes does? Have you lain on the couch for hours and even
days, trotting back and forth to the restroom every ten minutes
until you were too weak to get up anymore? Have you, in other
words, been through what I'm going through? For a few years I
wondered whether anyone else in the world could understand my
experiences.
    The stories I discovered on the Web taught me that, all things
considered, I was much better off than many Crohn's patients, who
often must deal simultaneously with a host of complications from
their symptoms (such as fistulae, which occur when the disease
burns a hole through the wall of the intestine and into a nearby
organ), with surgeries, and with related autoimmune disorders
(skin rashes, joint and muscle pains, other digestive problems).
They also helped me see that others did in fact have the same experiences as I did. As I will discuss in more detail in chapter eight,

those discoveries slowly helped me see my way to sharing my own story with others.

AFTER CLASS, reveling in the afterglow of seventy-five minutes of stimulating intellectual discussion, I realize again how profoundly this disease affects my thinking about just about everything. Four years ago I would have read *The End of the Affair* and not given one iota of thought to Greene's descriptions of illness.

I have lunch with a few fellow professors, and eat my regular meal in the cafeteria—a chicken wrap with mushrooms. I usually see any vegetables I consume in the toilet the following morning, but my system seems to tolerate mushrooms fairly well.

At 2:30 in the afternoon I do a reprise of the morning composition class, though without the close calls. This group of students is fun, and I enjoy myself.

I pick up a few things from my office, check my e-mail one last time, and head for home. My system has settled down by this point, and I have the luxury of a few hours without worrying about my bowels. Usually—at least during this current flare—I can feel comfortable that I will be able to keep things under control from around noon or so until seven or eight o'clock in the evening.

So when Katie and Madeleine clamor for a trip to the park a few blocks away, I agree to take them there for an hour or so before dinner. On the way down the hill over which our street passes we run into a neighbor playing outside with her children, and she decides to join us. The group of us walks together to the park.

Once there, I sit down with my neighbor for a few minutes on the bench while the girls attack the playground equipment with relish. My neighbor and I exchange pleasantries, talk about the children, about other happenings in the city. But we run out of pleasantries quickly enough—and I've never been much of a pleasantry person at any rate.

So in a few minutes I'm off the bench and I'm doing what I do best at the playground—playing. I start up a game of freeze tag, which the older kids on our street have just discovered, and volunteer to be "it" for the first time. A few other kids we don't know, supervised by bemused parents on the benches, join in.

Within a few minutes the air is filled with the squeals of mock-scared little girls, the thumping of feet on the play structure's suspended wooden bridge, and my mock-threatening growls. I chase the kids all around, just missing them at every turn, and I wonder if they realize I am letting them go, and it's all part of the fun, or do they really believe they are quicker than me?

I get lost in the play, in the pure enjoyment of body movement and the careful negotiation of the rules of the game, in the desire to help these small human beings experience these moments of pleasure as intensely as I do. Madeleine doesn't understand the game, she runs from me and squeals if I tag her, following the nearest older children to whatever point of safety they have managed to obtain.

Eventually I wear myself out and gather the girls away from their friends and the playground, and we make our way slowly back to the house, to Anne, to dinner, to the remains of the day.

AFTER DINNER I get a phone call from one of my neighbors, a child psychiatrist who lives three houses down and has a daughter the same age as Madeleine. He has tickets to a Celtics game, he tells me, which include a complimentary buffet dinner and free drinks beforehand inside the Fleet Center. All we have to do is sit through the lecture of some professor talking about a new drug for psychiatric illnesses. I learn later that this is a common practice. Drug companies sponsor events or dinners for practicing doctors to which they invite professors or researchers to come and speak about their products.

I am not an especially big fan of the NBA, but in general I enjoy live sporting events, and the free buffet dinner and beer make this invitation especially enticing.

But I hesitate nonetheless. The game is two weeks away, on a weeknight, and the Fleet Center—the home of the Boston Celt-ics—is forty-five miles from our house in Worcester.

"We're going to drive in to the train station," my neighbor tells me, "and take the train to the Fleet."

I work out the details. A forty-five-minute drive in rush-hour traffic, followed by another thirty-minute ride on the train. Repeat in reverse order for the trip home.

I wouldn't mind the drive so much if I were by myself, but with other people in the car it becomes complicated. Do I want to chance having to suddenly jump out of the car, in rush-hour traffic, and bolt to the side of the highway to relieve myself? The train presents an equally disturbing prospect—what if the urge over-comes me on a long stretch between stops? What if I have to jump off the train at an unfamiliar stop?

These sorts of calculations are so routine to me now that my mind runs through them automatically whenever I contemplate traveling anywhere.

In this case, I decide that my potential enjoyment of this expe-rience outweighs the possible risks, and I agree to go. As often as not, though, I will decline and elect to stay home, unwilling to take the chance of finding myself stuck somewhere, in a public place with strangers or even friends, without a restroom.

WE FINALLY BUNDLE the kids off to bed at around 8:00 PM, and Anne immediately changes into her pajamas and settles onto the couch for some television. She'll be asleep within an hour. Chasing around twenty-some seven-year-olds, and then coming home to chase our two- and four-year-old doesn't leave her much energy for weekday evenings.

Not that it matters much to me during the school year. I bolt immediately for the basement computer, and begin a marathon session of responding to the student drafts I received from my two composition classes earlier this morning. It takes me around fifteen minutes per draft, which means—with the occasional break—I can get to perhaps twelve of them this evening.

As 11:00 PM approaches, I can feel my concentration begin to sag, and I find myself surfing the Internet, checking a few Web sites on college football and reading up on the coming weekend's games. When I officially give up and log off a few minutes after eleven o'clock, I've only made it through ten papers. Twenty-one more will be waiting for me tomorrow morning.

I take my evening medications, pour myself a beer, and settle down to watch a rerun of *The Simpsons*. I drink two more beers before I make it to bed around midnight.

Beer, as my mother and wife have been reminding me for years, is no part of the ideal Crohn's disease diet, and I wouldn't recommend it to any fellow sufferer. In chapter nine I will describe in more detail the lesson I learned about the role that alcohol had been playing in my life, and the steps I took to modify that role.

But I haven't learned that lesson yet. So the last hour of my evening, in which I usually watch television and have a couple of beers—and have been doing so for the past ten years—represents the one part of my day when I absolutely don't worry about my health. Sitting on the couch, drinking beer, I feel like I'm in college again, before all of this started—healthy, normal, enjoying my physical self. Back to the days when I had never heard of Crohn's disease, and any diarrhea I experienced was just from the cafeteria food.

In bed a little after midnight, I look at the clock. Do I need to set the alarm? No teaching tomorrow. Anne will wake me up when she needs help with the girls.

REHEARSING THE ROUTINES of daily life reminds me how deeply the disease penetrates my existence—how much of my days are devoted to wondering where the next restroom might be, or worrying about how I'll be feeling in two weeks at the Celtics game, or in one year when I'm in Ireland, or thinking about any of the myriad effects of the disease in my life. If someone who could gain access to my brain offered me a minute-by-minute breakdown, I would ask them not to tell me what they had discovered. Learning how much of my life and my intellectual energies get nibbled away by the appetite of the disease would only depress me.

In the routines of daily life—interrupted, transformed, pocked, and staggered with medication schedules and trips to the restroom—the disease makes its presence felt most deeply. As I rehearse those routines, and lay them out on paper, I see how fully I am forced to build my life around the disease—how it determines and controls so many parts of my existence.

But I have learned that building my life around the disease beats the alternative: planning and living my life as if the disease didn't exist, and then suffering the consequences when it reacts and disrupts my life in ways I haven't anticipated. The disease may close off certain possibilities to me, but I would rather shut those doors myself at the outset than have them slammed on me when I have already stuck my nose in. And, in sometimes unexpected ways, as I will describe in later chapters, the disease may open other possibilities.

A life built around disease is still a life. Like almost any other life, the possible paths along which it might unfold are not endless, and are bounded by limitations and constraints. A clear-eyed, realistic understanding of the borders and obstacles of my path—an understanding that came for me only when I began to describe them on paper—may be the best tool I have to negotiate it successfully.

5

SMALL MIRACLES

Learning About God

[November–December 2000]

The Christ I read about in the Bible—a Bible I know from thirty years of Catholic masses, from Catholic school education from first grade through my M.A. degree, and from several personal readings of the New Testament during Lenten seasons—responds to human suffering and illness in a uniform manner: He heals the sick. Lepers, blind men, deaf men, epileptics, a woman with a skin affliction, men and women possessed by demons, Lazarus, the daughter of the Roman centurion, a paralytic, a man with a withered hand. When Christ meets suffering, he responds with healing.

He tells us, too, that we have only to believe in Him, and ask, and our prayers shall be answered: "Ask, and it will be given to you; search, and you will find; knock, and the door will be opened to you. For everyone who asks receives; everyone who searches

67

finds; everyone who knocks will have the door opened" (Matthew 7:7).

For five years now I have been knocking on that door; I'm still waiting for it to open.

IT IS A COLD DECEMBER DAY, at the edge of the semester's close, and I lock my office door and trek out into the lightly blowing snow, across the campus toward the college chapel. The walk isn't long, perhaps a few hundred yards. Still, I take it slowly, testing whether the walk will loosen up my system and send me sprinting back to the bathrooms in my office building. The chapel has no bathroom in it; I asked someone, of course, the first time I attended a mass there. I always ask.

The chapel is warm and welcoming. Soft lighting lies upon the long, arched wooden beams stretching up to form the roof of the building. Behind me the rays from a clouded winter sun filter through the stained-glass rear window of the chapel.

It is quiet enough in the chapel that the smallest rustlings, the most discreet coughs, are amplified throughout the large and near-empty structure. An elderly gentleman—I've seen him at every noon mass I've attended—greeted me when I came into the back of the chapel. Scattered throughout the pews are perhaps eight other mass-goers: most of them elderly, from the neighborhood around the college. I see one middle-aged woman who might be faculty or staff. No students.

Next to the altar the priest sits quietly, dressed in the purple robes of Advent, his hands folded in his lap. He seems absorbed in thought, or perhaps prayer.

The bell outside the chapel rings twelve times, and the priest stands and raises his arms, signaling for the dozen of us in the pews to follow his lead.

Mass has begun.

SINCE THANKSGIVING, when we had spent five days in St. Louis visiting Anne's parents, my illness had been steadily worsening. From four or five bowel movements a day through most of the fall, I was edging up to seven or eight, with spots of blood here and there, and an increasing number of close calls. I was staying at home or in my office as much as possible, venturing away from familiar restrooms only if it was absolutely necessary. Worst of all, I could feel myself just beginning the slow slide into depression that had overtaken me in August.

I had been trying to taper off prednisone throughout the fall semester and it hadn't been working. I could reduce from forty to thirty to twenty milligrams without much trouble, but every time I dipped below twenty milligrams per day the disease became active almost immediately. I would ratchet the dosage back up until I settled down and then try again. In late November I went through the same cycle, but this time, when I ratcheted myself back up to the higher dosage, I didn't settle down.

I was a little smarter in late November than I had been in August, and in early December I made an appointment with the gastroenterologist I had begun seeing after I switched from the university doctors. I was hopeful that an earlier intervention might produce better results.

The first time I had met with this new gastroenterologist, in October, we had discussed the possibility of my starting Imuran, an immunosuppressive medication that was becoming increasingly common for Crohn's disease treatment. Imuran had been developed initially for kidney transplant patients. It suppressed their immune systems after surgery, preventing their bodies from attacking the foreign organ. In much smaller dosages, the drug had been tested on patients with Crohn's disease, and had shown remarkable results. I had heard reports of patients with disease patterns similar to mine, or worse, enjoying remissions for as long as four or five years.

The university doctors had strongly recommended that I begin taking this drug, as had my gastroenterologist in Chicago during my last visit with him in the spring. I had demurred on those occasions, for the not-very-good reason that I didn't want to add another drug to the small pile of daily pills I was already taking. I did not think through the prospect enough to realize that, should the Imuran work for me, it would allow me to reduce or eliminate some of those other medications.

I had learned along the way that Imuran also has helped taper patients off steroids. So by the time I saw my new GI in October, I was ready, and I told him so.

"I'd like to hold off on that for a little bit," he said to me initially, to my surprise.

"Why?"

"Since I have not seen you before, I want to see how you respond to the more conventional medicines first, and get a sense of your disease pattern. I know that Imuran is becoming more popular as a first-line medication for Crohn's patients, but I view it with caution. This is a serious drug."

"What do you mean?"

"I mean that we are still learning about its potential side effects. There have been some cases of lymphoma associated with the disease. It can render you more susceptible to infections and contagious diseases. And are you planning on having more children?"

We were.

"There may be risks of birth defects."

I was taken aback by this, and must have shown it.

"I don't want to scare you," he continued. "The number of

cases of lymphoma is rare enough that they have been written up in medical journals. And we don't really know for certain about the effects of the drug on the reproductive system. But I want us to approach it with caution. It is a serious drug, and we should use it in serious circumstances. I am not convinced you're there yet."

So we had agreed to wait. But as the semester drew to a close, and my condition began to worsen gradually, I had been rethinking that decision. When I made the appointment with him in December, I was certain that I should begin the new drug.

But on the day I went into his office I was having a good morning, feeling stronger than I had been recently and my bowels slightly more under control. Even though I know perfectly well that one can't make any judgments about the disease based on a single day, I can't help but be affected by especially good and bad days.

So in the doctor's office in early December I hemmed and hawed, and he hemmed and hawed as well. Finally we agreed that, with four weeks of semester break coming up, I would have time to relax, to sleep, and to heal. We would see if I improved over break; if I didn't, we would begin the new medication in January.

That good day—of course—had been followed by a bad one, and I had not been improving at all in the days since my appointment.

IN THE CHAPEL I was settling into the routines and rhythms of the mass: the welcoming, the opening prayers, and then the readings of the gospels. For some silly reason, I had an expectation that God would speak to me through the Gospel reading that day—that it would be a story of suffering and healing, meant to comfort me.

It wasn't. I can't even recall what the reading was about anymore, but I remember feeling disappointed that it didn't apply to me. I listened to it, and to the brief sermon of the priest that followed. In that sermon, I waited for a special message from God, through the mouth of the priest. Again, nothing.

I have never been much enamored of the mechanics of the Roman Catholic liturgy. I know Catholics who find the mass an inspiring and aesthetically pleasing spectacle; they kneel devoutly and ecstatically before the risen host at the celebration of the Eucharist, their hearts lifted by the thought that Christ's body will enter their own.

I am not one of those people. I have always struggled through mass. My mind wanders, my knees and butt get sore, and I spend much of my time watching the activities of my children, or of other small children around me. I generally listen to the Gospels, and the homily, and find material for reflection in those parts of the mass, but I spend very little time in mass praying in the way that I suspect those around me are doing.

I go, though. I've been going since I was born. I skip mass a little more frequently these days than I did when my parents herded us all there every Sunday of my childhood, but even now, more often that not, Sunday mornings will find me in church.

I am reminded of my long and familiar association with the mass as the priest recites parts of the mass that I have—without thought, through sheer volume of repetitions—committed to memory:

"Lord Jesus Christ, you said to your apostles, my peace I give to you, I give you my peace. . . . Look not on our sins but on the faith of your church as we wait in joyful hope for the coming of our savior, Jesus Christ. . . . Let us all share with one another a sign of Christ's peace. . . ."

At this signal to greet my neighbors in the pews and offer them the sign of peace, I look around rather futilely at the empty rows separating me from the nearest human being in the chapel. I receive a few nods and smiles from afar, and flash a two-fingered peace sign to a woman across the aisle from me.

This brief interlude of social interaction within the mass concluded, the priest begins the rituals of blessing and preparing for the celebration of the Eucharist.

WHEN I REFLECT upon the literally thousands of masses I attended while I lived in my parents' home, what stands out for me are feelings of warmth and comfort, both of the familial and spiritual kind.

No doubt I have telescoped the recollections of many years of masses into a few short minutes of videotaped memory, but the images that I recall most clearly are those associated with the ritual of the sign of peace—that moment in the Catholic mass when we are told to turn to our neighbor and share with him or her some sign of peace.

In my family that meant shaking hands with our siblings, with whom we were often—as siblings often are—quarreling. But no matter how serious the quarrel, no matter how irritated we were with one another entering the church, we were capable of forgiveness in that moment. The particular image that stands out in my mind from those family masses are the rueful smiles we would exchange, along with our handshakes: Wasn't I being an idiot this morning? And weren't you being one too?

The weekly ritual of mass attendance came to represent family. It was the only time during the week when all seven of us were guaranteed to be together in the same place, with none of us running in or out of the door on the way to football practice, or gymnastics, or school, or an evening out. Especially as the older siblings grew into their high school and college years, those weekly gatherings of the family became one of the few guarantees we had of seeing one another, all together.

As the sign of peace concludes, my mind drifts to thoughts of my family, and to the fact that in just a few short weeks almost all of them—with the exception of my eldest brother Tom, his wife, and their three children—will be gathering at our house for a few days after Christmas. My sister and her new husband are driving in from Chicago; my parents are doing the same from Cleveland; my youngest brother and his new wife are driving up from New York City; and my older brother Tony will be driving in with his wife and their son from New Haven.

We will be packed into our four-bedroom house, with at least one couple on the sleeper sofa in the family room, but I am certain no one will mind the crowding. Crowding is what our family has always been about, with five children and two parents growing up in a four-bedroom home in the west suburbs of Cleveland. Until Tony left for college, I never had a bedroom of my own. At one point during my childhood, after the birth of our youngest brother, I shared a bedroom with my two older brothers: two bunks crammed into one corner of the room, another bed across from it, and desks and dressers wherever we could find space for them.

I have been anticipating these few days of our holiday family reunion for months now, and can hardly stand the wait for another couple of weeks. When we are together we revert to the habits of our youth, from exchanging bone-crushing bear hugs to staying up late into the night, drinking beer and renting classic comedies— anything with Bill Murray will do (except *The Razor's Edge*, of course).

What I don't know yet, sitting in the chapel and letting my mind drift forward in anticipation, is that by the time those few days roll around I will be even sicker than I am at that moment. I will be in the bathroom as many as ten to fifteen times a day, I will be eating bland food to no avail, I will be so tired that I will occasionally have to leave the laughter and music in the family room for a short rest upstairs. I will be a source of worry and concern to all of them—but especially, again, to my mother.

What I don't know is that on the Saturday afternoon after Christmas my father will buy loads of shrimp cocktail from the store, that my little brother Billy and his wife will make their famous guacamole dip, that Tony will open a bottle of wine long before the customary cocktail hour, and that we will all gather into the family room, eating and drinking and playing with the children for several happy hours. I don't know yet that I will be too sick for any of it, that I will sit on the floor over by the stereo, changing the CDs, running back and forth to the bathroom, drinking the glass of water I have begun to carry with me everywhere to stave off the dehydration—and watching.

What I also don't know is that, over the course of those few hours, and over the course of that visit, I will begin to understand that my happiness does not necessarily have to depend upon my health. I will feel happy despite my body's best efforts to thwart me, despite the fact that I will be the only one not gorging myself on shrimp, and not drinking. Even my mother, a near teetotaler, will sip a glass of wine, her face flushed and warm.

This understanding will not undercut or alter the lessons I have been learning about the disease's influence on my mind and my emotions as well as my body. But I will begin to see that it is possible, even in the darkest moments of disease, to find solace and comfort in those parts of my life that have always provided solace and comfort to me: my family—both the one from which I have come and the one I have created with my wife and children—and my religion.

Sitting in the chapel on that early afternoon in December, I don't yet know all of this. But, in the course of that noon mass, as I follow the familiar and comfortable rituals of this religious ceremony I have been attending for thirty years, I have begun to sense it.

We are now in the celebration of the Eucharist, and we stand—not kneeling, as we did at the church in which I grew up—and watch the priest perform the ceremonies that, according to Catholic doctrine, will turn the pieces of bread and cup of wine on the altar into the literal body and blood of Jesus Christ.

I have always felt conflicted about this doctrine of Catholicism, which requires me to suspend my belief in the laws of physics and biology in order to accept the occurrence of this miraculous transformation at every mass I attend. I have difficulty believing anything I cannot understand rationally, and this piece of Catholicism ranks near the top of doctrines that are rationally incomprehensible. So for most of my life I have received the Eucharist, and watched the rituals associated with it, with some dispassion.

But a few years into my struggle with the disease, while I was sitting in a mass one Sunday morning, it occurred to me that the Eucharist presented an opportunity for a sort of divine intervention in the course of my disease. It made perfect sense: Here was a piece of divinity that, instead of existing in some faraway realm of angels of heavenly bliss, actually passed through my gut! Was it possible that God could use the Eucharist to soothe and heal my gut, if only I were to believe that He could do so?

The idea quickly took hold of my imagination. We were still living in Chicago then, and the disease was in a semi-active state. I could live my life, and was not experiencing any of the emotional symptoms that accompanied a serious flare, but diarrhea, the occasional close call, and a persistent, low-grade fatigue were all parts of my routine.

At that time we attended mass at a church that held a separate celebration, in the basement of the church, for parents with small children. It was packed with people, and with fidgety, loud children—you had no worries whatsoever that your children were disturbing anyone. In fact, you could hardly hear your own kids for the noise from other people's children.

I liked this aspect of the mass, because I have always felt that the mass should be an occasion for laughter and happiness rather than solemnity. The presence of the children necessarily made it so.

But when I began to think about the healing power of the Eucharist in my life, I began to resent all of the commotion and distraction. I wanted to pray and contemplate the Eucharistic celebration in silence, and to have a solemn atmosphere in which I could communicate to God that I really believed He could heal me.

I felt certain that this was the trick: God had obviously chosen not to heal me through the Eucharist thus far, which meant that there must still be some contribution I had to make to the process. I understood that contribution to be my complete faith in the possibility that God could heal me in this way.

I probably did not articulate this to myself at the time, but in retrospect I suspect this belief came from various biblical stories in which the belief of the sick person seems to play a crucial role in Christ's decision to heal: the story of the Roman centurion, for example, who begs Jesus to heal his sick servant, but then balks when Jesus tries to follow him to his home. "Sir," the centurion replies, "I am not worthy to have you under my roof. Just give the word and my servant will be cured." Astonished, Jesus announces to his disciples that "in no one in Israel have I found faith as great as this." The sick servant is cured.

So I needed to become that centurion, so certain that Christ had the power to heal me that He would finally consent to do so.

And I felt certain that that healing would come through the Eucharist. That little sliver of unleavened bread, I was convinced, would work its way into my colon and soothe and heal my sores in a way that conventional medicine never could.

So I prayed. And I went to mass and received the Eucharist. I did my very best to believe in its healing powers, blocking out the sounds of children all around me as I sat in church and felt it sliding into my digestive tract. I tried to envision it as a little circle of healing light, one that would illuminate the dark spaces of my colon, bearing the divine cure to me.

And—of course?—nothing happened. Did I really expect it to? I'm not sure anymore. Perhaps I never really managed to dispel my doubt that I could be healed in this way, and perhaps that sliver of doubt prevented me from receiving the sort of healing that I hoped would come from my absolute faith in Christ's powers.

My eventual abandonment of the hopes I had for such divine intervention were slow to die—so slow, in fact, that even during this December mass in the chapel I am still grasping at shards of that hope. I never made a conscious decision of any sort to reject the possibility of such healing; instead, as the Sundays stretched into months and then years, without any consistent improvement in my health, I resigned myself—or re-resigned myself, more accurately—to the idea that I could not count on God to solve my problems for me.

So as I walk slowly down the aisle to receive the Eucharist, I am reminded of my hopes for divine healing, and feel the tiniest surge of hope. Perhaps this time?

I receive the bread, place it in mouth, and say a little prayer.

But even as I do so, running beneath that prayer like a river, another part of my brain is chastising me for the falseness of it; I don't really believe it. That surge of hope is reflexive, unthinking; every other part of my brain understands that healing will not come to me in this way.

To the unbeliever, and even to some believers, this long spell of belief I had in the possibility of divine healing will seem foolish. But it is an old truth that there are no unbelievers in a foxhole; I doubt there are many unbelievers, either, among the chronically and terminally ill.

Many of those chronically ill believers look to the divine, I know from both intuition and experience, with the same hopes that I have spent the last few years struggling to overcome. They believe their deity will, out of compassion for their plight, perform a miraculous intervention in their lives and render them whole again.

But I hope that at least some of them look to the divine, and to the religions that formalize their relationships with the divine, with the understanding that has been gradually dawning on me throughout this noon mass, and that will become so important to me in the months of illness that will follow.

MY FIRST ARTICULATE SENSE of this understanding comes to me as I am back in my seat in the chapel, watching and hearing the familiar closing rituals of the mass. With a start, I realize that, for the past half hour or so, I have been happy.

I have been happy to be in a place so long associated for me with comfortable and familiar rituals. I have been happy to be in a place in which I can take the time to reflect in an unhurried manner on my life, on God, and on the relationship between the two. I have been happy to be in a place in which I feel a connection,

however vague and fuzzy it might be, to a community, a world, and
a spirituality larger than myself and my immediate concerns.

I have been happy to stand among flawed human beings striving
to be better, all of us affirming together the importance of human
community, if only for just thirty minutes on a cold December
afternoon.

MASS HAS ENDED, but I am reluctant to leave. I linger in the
pews, reveling in this newly articulated awareness and understanding of the role that God and religion should be playing in my life.
Outside, the snow has been coming down more insistently, the
wind blowing more coldly. Outside, there are papers to grade, students to see, classes to prepare for. Inside, there is peace, calm,
comfort.

Over the next days and weeks, as the semester comes to a
close, as I continue to anticipate the arrival of my family, and as
my disease worsens, the insight that flashed upon me in the chapel
that afternoon will work its way through my understanding of every
part of my spiritual and religious life.

That insight will help me see the value of my God and my religion in providing me with comfort and peace even in the worst
stages of my disease; it will help me see the value of the mass as
providing me with familiar rituals in times when I can count on
almost nothing else in my life to remain constant, from the plans I
make and am forced to break to the number of times I visit the
bathroom each day; it will help me understand the value of prayer
as an end in itself, rather than as a means of securing favors or
blessings from God.

It is this last point that has perhaps become most important
to me.

The period during which I prayed for Christ to heal me

through the Eucharist, while it may have been unique in its concentration upon that one sacrament, was not out of character for me—or, I suspect, for many sufferers of chronic disease—in other ways. For most of my life, even before chronic illness, and like most people raised in religious traditions, I have been accustomed to pray in moments of crisis.

For the past half-dozen years, those moments and periods of crisis primarily have concerned my illness. Prayers for general healing, for a respite from illness in the darkest days of severe flares, and for, most simply, a handily placed public toilet in a moment of urgent need—I have let all of these prayers issue from my lips at one time or another.

And, unlike those blessed characters from the Bible with whom Christ came into contact, my prayers have almost uniformly gone unanswered. Underlying my sense that God could intervene in the course of my illness and heal me, I can see now, was the belief that God had given me Crohn's disease for some reason I could not fathom.

It has taken me almost five years to realize that God did not give me Crohn's disease—and, more importantly, that God is not going to take it away. The process of evolution, of genetics, and of environmental conditions gave me Crohn's disease. I am quite certain that God is sorry to see me suffer, and mourns with me when I mourn. But I'm also quite certain He has no plans to intervene in the course of my disease.

Does God work small miracles in people's lives—using the Eucharist, for example, to heal their digestive disorders? I don't really know. I don't know how anyone could ever know that. I hope He does, for the people who really need Him. But my disease has taught me that I cannot rely on such miracles to get me through my problems, and that my prayers are best aimed in other directions.

I pray now for the strength to fight whatever the disease can throw at me, for the wisdom that disease can bring, and for the patience to see me through to the next remission. Those are prayers that, once uttered, are already half-answered. The process of asking God for strength helps steel one for the battle; the process of asking for wisdom brings reflection and understanding; the process of asking for patience helps see one through another day.

In the movie *Shadowlands*, the character of C.S. Lewis explains at one point that he prays not to change God's mind, but to change his own. For some reason this line stuck in my head when I first heard it, many years ago, but only in the last year or so have I come to understand what it means. I pray to heal my mind, not to heal my body. And the very act of praying begins, and sometimes completes, that healing.

These spiritual insights are perhaps the most profound lessons that chronic disease has taught me. Look to what the very act of praying can do for me; don't look for answers. Go to mass to see myself as part of a community and a religion that help to place my individual problems in perspective; don't go to mass expecting miracles. Go to mass to affirm my place in this religious family.

I would be dishonest if I did not end this chapter by confessing, though, that I don't believe I can ever shut myself off completely from the possibility that answers may come, that a miracle cure for disease might be right around the corner, and that God may have a different message for me about religion and prayer.

My mother, I know, would insist that such miracle cures are entirely possible, and in fact occur quite frequently. I know she prays for such healing for me, and I know she attributes her own recovery from cancer to just such divine intervention.

But I will take, in the place of such healing, the next time she and I are able to attend mass together. The ritual of the sign of peace will come around, and she and I will exchange hugs. She will smile and look into my eyes, and she will squeeze my hand.

Such little gestures: miracles of their own.

6

THE HOSPITAL

Learning to Hit the Bottom of the Well

~

[February 2001]

In January of 2001, my gastroenterologist finally prescribed the immunosuppressive Imuran, in the hope that the drug would help me taper off steroids and achieve a more sustained remission. But the combination of the prednisone—which also helps suppress the immune system—and the immunosuppressive left me vulnerable to infections and contagious diseases, and in mid-February I came down with a nasty stomach flu.

It happened on a Monday, just after dinner. Anne had taken the girls out shopping for something or another, and I had a quick dinner alone. Afterward I brought in some firewood from outside, intending to make us a fire for the evening. Outside it was dark, and had been since 5:00 PM, and we were in the midst of a stretch of frigid weather that had yet to dump any snow on us.

Before I had a chance to make the fire I began vomiting and could not stop. The disease had been more than usually active that day, and within a few hours after the vomiting had begun I was living in the bathroom, everything in my system coming out of both ends at regular intervals.

The vomiting and the diarrhea slowly worsened, and it wasn't more than five or six hours after I first became ill—between 11:00 and midnight—that I realized I was becoming dangerously dehydrated. We called a neighbor to watch the kids, my wife drove me to the hospital, and I spent the night in the emergency room, receiving IV hydration.

I wanted to be released in the morning. I didn't see that the hospital could do much for me. There is no cure for the flu, of course, and the diarrhea was the result of disease activity we all knew about. I was already taking all of the medications I could possibly take, so I didn't see the point in sticking around and losing sleep in the noisy emergency room.

They did release me, and early that morning—before 7:00—I called Anne and asked her to pick me up before she left for school. As I waited in the lobby for her, I suddenly began to feel sick again. I ran to the bathroom, where I vomited into the trash can while I sat on the toilet.

I should have turned around and walked back into the ER, but I didn't. It still seemed that the hospital had nothing to offer me.

It didn't matter much in the end. I was back in the ER the following evening, the vomiting less serious but the disease escalating out of control. When I had described the full extent of my symptoms to the nurse on call earlier that evening, she had sent an ambulance for me, so at least Anne and the girls were able to remain in bed—though I don't suppose Anne was sleeping much. I spent another night in the hospital, during which I had a very strange encounter with the nurse who had been assigned to my bed the previous evening.

THIS NURSE WAS NOT responsible for my room on that second evening, but she came in to speak with me anyway. As she talked she busied herself in the room, straightening up medical equipment, folding linen, and adjusting my bed and pillows.

She came by, she explained, because she wanted to talk to me about my illness. Her father had Crohn's disease, and for many years he had suffered as I apparently seemed to be suffering—in and out of hospitals, bouncing up and down on multiple medications, and experiencing all of the disease's unpleasant symptoms and effects.

At that point in her story I expected her to offer me the customary words of emotional support and sympathy.

It turned out, though, that she intended to offer me a more specific message.

"He finally got better," she said, referring to her father, "when he got off all of the medications they had him on and changed his diet. He went on an all-juice diet, and that cleared up his inflammation. Then he stopped taking his medications and began to eat right. He hasn't been back to the hospital since."

"Is that right?" I said.

"Listen," she told me, lowering her voice to a conspiratorial whisper. "If you keep coming back here, all they're going to do is try and put you on more and more medications. And the more medications they put you on, the more you'll come back here. It's a vicious cycle.

"They want to help you, but you don't *need* this kind of help. You need to take control of your life, take control of your diet, take control of this disease. You're the only one that can make yourself better."

She grabbed a pad of paper and wrote down several names on it.

"These are books that will help you," she said, still whispering. She thrust the paper into my hand. "These books will tell you what to eat, and how to get away from all of these medications."

She looked at me intently, still pressing the paper into the palm of my hand.

"Do you promise me that you will read these books, that you will start taking control of this disease on your own? Do you *promise me* that you will?"

I almost burst out laughing at the absurdity of it. Sitting there on the edge of my hospital bed, in her nurse's aqua-green uniform, stethoscope around the back of her neck, she was asking me to renounce everything that all of those trappings stood for. Undoubtedly she would be leaving my room to visit other patients—to administer their medications, to monitor their vital signs with expensive medical equipment, to follow the dictates of the emergency room doctors in the care of her patients.

"Sure," I said finally. "Sure. I will."

She gave my hand a squeeze and left without saying anything else.

IN THE MORNING I requested, once again, to be released. The emergency room doctor asked me if I thought I needed to be admitted. I told him I understood what was happening, and didn't see any reason for it. He agreed.

Anne was staying in regular contact with my mother, who decided to fly out and help us deal with all of the other aspects of life—like taking care of children and running a household—that the disease was eclipsing. She brought with her everything that a mother brings: love, compassion, help, and—especially—worry.

But she could not bring me help for my inflamed and ulcerated gut. Even as the flu abated, the disease began to spiral out of control. By the time I was finally admitted to the hospital two days later, I was having between twenty-five and thirty bowel movements in a twenty-four-hour period.

The nurse who greeted me when they wheeled me from the emergency room into the room where I would stay for the next six days asked me about this in disbelief:

"Are you the guy who was having *thirty* bowel movements a day?"

"That's me."

"Is that *really* true?"

"I wouldn't make it up, believe me."

To discover that my symptoms were shocking even to medical personnel certainly let me know, if my body had not yet done so sufficiently, that I was in serious trouble.

Once I had been admitted and settled into my room, they hooked me up to an IV machine that dispensed extremely high doses of prednisone into my system, along with fluids to combat the dehydration. I was in a room far at the end of the hall, where—as one nurse kindly explained to me—I wouldn't be too disturbed by foot traffic or other noise outside in the hall. Anne packed the few clothes I had brought with me into the dresser drawers the hospital provided. She sat with me for a few minutes, but we both agreed her time was best spent at home with the girls.

Once I was alone, I began to worry immediately. Not, as one might expect, about my health—I was worried, instead, about school.

I had missed almost an entire week of classes, from Tuesday

through Friday, and it was clear that I would miss at least one more week. If the amount of time I had to miss were to stretch into three and four weeks, the college would be better off giving me the rest of the semester on leave, and hiring someone to take over my courses.

I did *not* want to do that, primarily because I knew we could not afford it. So I began thinking immediately about how I could have my classes covered for the following week without overtaxing any of my colleagues in the department. Assuming I would lose the next week in the hospital, I just needed to make it back for the week following that one, and then it would be spring break. I could use spring break to recover and regain my strength.

My first visitor on that first day in the hospital was the chair of my department, who absolutely assured me that the department would cover the classes until I could return, and that I should worry about nothing but my health. I wish I could have listened to that message.

Her visit simply stoked my school anxiety further. Although I should have been resting, I couldn't sleep—the prednisone, which induces sleeplessness, was at least partially responsible for that—so I plugged in my laptop and typed a long message to send around to everyone in the department, requesting their assistance. I saved the file onto a disk and later, when another friend from the department visited me, I asked him to give the disk to the department secretary and have her e-mail it to all of my colleagues.

Reprinting that message here will perhaps give the clearest picture of how I could not—despite everything that was happening to my body, mind, and family—accept the idea that the world would go on spinning without me:

Dear Colleagues:
    First of all, thanks to all of you for the flowers, and to many of you for your individual expressions of concern and offers to

help. And especially thanks to Mike Land for helping me coordinate all of this, and thanks in advance to any of you who can
respond to this call for assistance. . . .

I canceled classes last week from Tuesday through Friday,
but this week would like to try and have most of my classes meet
if that will be possible. The great news is that I had planned this
week to have almost no preparation, because I needed to write
a conference paper this week! I have canceled that conference,
so that means I can relax this week. To add to the good fortune,
my mother has come from Florida to help out around the
house, and my wife has off next week because of school vacation. So she can help do some of the leg work for this stuff.
Having my Mom here is a weird kind of irony, because this place
reminds me of Florida, when we visit my parents on spring
break—lots of old people, not much to do besides read and
relax. Of course the beach and restaurants would be nice, but
you can't have everything.

At any rate, you will see from below that in almost all of the
classes I have to teach this week, with one exception, absolutely
no preparation is required at all. What I hoped to do was lay
out the dates and times for all of my classes, and if you are able
to perform some very minimal duties you can let Mike know
and then just give me a quick call at the hospital if you have any
questions. My number here in this palatial suite is 363-8649. I
am pumped up on IV steroids so I am ALWAYS awake. So here
goes. . . .

The message went on to describe what I was doing in each of
my classes, with detailed instructions for my substitutes.

Despite the good humor I mustered in parts of the message,
the emotion that animated these instructions was panic. When I
had been sick in August, it had been prior to the start of school,
and the lack of defined and immediate tasks had allowed me to slip
into that state of apathy and depression that lasted me into the

beginning of the school year. I wasn't getting anything done then, but it didn't really matter.

This time around, I had a schedule—courses to teach, meetings to attend, students to advise, and writing deadlines to meet. Moreover, at our college we teach on a 3–4 schedule, which means we have three classes during one semester and four the next. The spring semester was my four-course semester, increasing my duties by 25 percent. Finally, the rapidity and severity with which the disease had overtaken me this time meant that I did not have weeks or months to spend slowly sinking into the usual state of apathy or depression that accompanies a longer flare. When I entered the hospital I was still in the mental and emotional state that usually lasts me through the first days and weeks of a flare—trying to fight off the disease, to postpone the inevitable by exercising more regularly, watching my diet more closely, and getting extra sleep.

This mode always carries with it a great deal of anxiety and frustration, because such homeopathic therapies never work. If I have reached the stage at which I am aware that a flare is coming on quickly, then I am usually too far gone to help myself without more conventional medicines.

This mind-set—the attitude that I can fight off the disease, and will not let it interrupt my life—is probably more responsible than any of my other behaviors for the severity of some of the flares I have experienced. It has led to the postponement of regular medicinal interventions, to higher levels of anxiety and stress, and to a divided mental state in which I am desperately trying to ward off the disease on the one hand, and increasingly frustrated, on the other hand, at the ineffectiveness of my efforts.

More than anything else, the hospital helped me to see the futility of this attitude and the toll on emotional, physical, and psychological energy that comes with it. My stay in the hospital helped

me learn how to hit bottom and let go—to learn when to surrender myself to the therapies and medications that will help me
recover from a flare most effectively and most quickly, and to let
modern medicine do its work.

That sort of letting go does not mean abandoning control of
my medical care—as I will explain shortly, my continued monitoring and input on that care remained essential. But it does mean
that, when I have really hit bottom, I cannot expect my body to
heal itself miraculously, or to respond quickly and easily to the
sorts of suggestions that the nurse had given me on that second
night in the emergency room: all-juice diets, or radical dietary prescriptions of any sort.

I remain hopeful that those sorts of remedies might one day be
effective for me in maintaining remission, or in managing moderate
disease activity. But when I am lying at the bottom of the well, the
recommendations of that nurse are a siren's song: They tempt and
tease me with the belief that I can take control of my disease and
heal myself.

Letting go, by contrast, means acknowledging that the disease
has temporarily taken control of my life, and that I need help to
regain that control. Letting go means accepting that help willingly
and gratefully.

MY SIX DAYS in the hospital settled quickly into routine. Of
course, routines come easily enough when you have an IV line
attached to your arm and can't venture more than fifty feet from a
restroom.

Someone usually woke me up around 6:30 or 7:00 AM to check
my vital signs and give me some medication. After that initial wake-
up, I would lie in bed for another hour, drifting in and out of sleep
until breakfast.

When breakfast was finished at 8:30 or so, I read until mid-morning, when I usually tried to sleep a little. After lunch I did the same thing. Read for a couple of hours, and sleep for a couple of hours.

Anne usually arrived around or shortly after dinnertime, and we would sit together and read our books or watch television. Sometimes we lay together in my hospital bed; sometimes I would let her lie in the hospital bed and I would sit in the chair in my room, just for a change.

Usually Anne came alone. She had brought Katie and Madeleine on my first evening in the hospital, but it was quickly apparent to both of us that my hospital room was not very child-friendly. After fifteen minutes of hugging, playing with the controls on my hospital bed, and peering out at the hallway traffic, they were ready to go. We agreed, after that initial visit, that Anne would leave the children with my mother in the future.

Anne usually stayed until nine o'clock or even ten on some evenings, and when she left I watched another hour of television or read. The night nurse came around 10:00 with my final medications for the day, including sleeping pills. I took those at 11:00, and was usually out for the night by 11:30.

All of this was punctuated by regular trips to the restroom, though the daily number of those trips began gradually to decrease: from twenty-five or thirty to twenty, to fifteen, and down to eight or ten by the time I was ready to leave.

As the week passed, I began slowly to loosen my hold on the world, and to learn to let go in a second way: to relax my desperate grip on those parts of my life that would survive, at least temporarily, without me.

What I needed was simply to rest. I had initially asked Anne to bring my laptop and all of my schoolwork to my room, so that I could read ahead for when I returned. She also brought me, though, Arthur Golden's *Memoirs of a Geisha*, a novel she had recently read and loved. For the first day or two I worked half-heartedly at my schoolwork. By Monday I had abandoned it and was spending all of my time reading—and enjoying immensely—Golden's novel.

I had a massive stack of papers to grade; they sat ungraded in my schoolbag in the closet. I had lots of phone calls I wanted to make, to friends and even family members with whom I had not spoken in some time; I enjoyed the quiet and silence of my room. As usual, I had essays and children's books and novel ideas floating around in my head, waiting to be put down on paper; my laptop remained in its carrying case in the corner of the room.

The farther along in the week I got, the more I began to sleep and relax, and slowly, almost infinitesimally at first, I could feel my body beginning to heal.

IN THE HOSPITAL I did discover one area where I could not relax my guard, however.

I had initially brought all of my medications to the hospital with me when we came for the third and final time. The nurses let me know immediately, though, that the hospital would take control of my medication schedule while I was in their care, including both providing my pills and bringing them to me at the appropriate times.

This worked fine until my third day.

As always, I had one medication I was supposed to take at around 2:00 PM. This was also usually the time that the nurses changed shifts. The afternoon nurse generally would come in to see me with my medications and introduce herself sometime between 2:00 and 2:30.

On that third day, I had slept for a while after lunch, and woke up to begin reading around 1:30. I had one eye on the clock, waiting for my early afternoon medication. When you have so little to do and think about, each visit from the nurses becomes a marker in the passing day. Two o'clock passed without a visit, as did 2:30 and 3:00. I continued to read halfheartedly, expecting a nurse to come swirling in at any moment, in a rush from some emergency in another part of the hospital, apologizing for her lateness.

By four o'clock I suspected that someone had overlooked me, and I buzzed for a nurse.

"Hello," she said brightly as she entered my room, a few moments later. "Did you buzz?"

"Ahhhhh . . . In the past, they had been giving me my Asacol at 2:00 PM, but I didn't get any today. Did they change my schedule?" I didn't want to accuse her directly of neglect.

"Oh," she said, genuinely perplexed. "Hold on a second."

She returned five minutes later with a chart in her hand, and a small cup with my pills in it. "I'm sorry," she said. "Somehow we got this mixed up."

Of course human error occurs in every profession, and of course this particular mistake was quickly rectified and, in the big picture, probably would not have made any difference in the course of my healing. But it was disconcerting for me to see it happen, and to be reminded that human error was as likely to occur in a profession in which people's health and lives were at stake as it was in any other profession.

I suppose I should have been prepared for something like that to happen, had I thought more about the food they were serving me.

The doctor who saw me on my first full day had suggested to me, in a general conversation about my health and diet, that I avoid lactose for a while. Lactose is difficult to digest, and so—even if I were not lactose-intolerant—it would make sense for me to keep it out of my diet until I had settled down.

On my very next meal tray, I received a carton of milk.

I left it untouched, thinking that perhaps the doctor's orders had not reached the food service people yet. At the next meal, and every meal thereafter, a carton of milk showed up on my tray.

For the first few days I was there, they kept me on a liquid diet of broth, Jell-O, juice, and water. When they finally switched me off that diet, I was expecting to receive softened and easily digestible foods like rice, pasta, and whipped potatoes. Instead, the first tray of regular food I received included a cup of raw fruit, green beans, and chicken.

I had enough experience with dietitians to know that raw fruits and vegetables were probably the worst substances I could put into my body at that point. Sending fruits and vegetables into my damaged colon was, as one dietitian colorfully described it to me, like running a whisk broom over an open sore.

I picked away at the pasta side dish, drank the juice, and ate a few pieces of chicken. It was not the sort of first solid meal I had been expecting.

After a day or two more of this I began to get a little card with the daily menus on it, and I could select my own foods. I wondered if anyone was monitoring that. If I selected bran flakes and milk, fruit, and raw broccoli for every meal, would they bring it to me? In retrospect, it seems highly doubtful that anyone would have said anything.

Eventually I said something to the nurse and they switched my thrice-daily carton of milk to one of Lactaid, a lactose-free milk that tastes virtually the same as milk.

I had one final experience with the doctors in the hospital that reaffirmed the commitment I had made earlier that year to becoming my own best patient advocate.

The first physican who saw me was a general practitioner, a young man with a friendly bedside manner who apparently did the rounds covering my ward.

He, to the best of my understanding, ordered my dose of IV medication and controlled and monitored any other treatment I was to receive.

On my first day in the hospital he laid out a general plan of treatment for me. I would remain on the IV steroids for a day or two, and would then begin taking oral steroids at a dose of 80 milligrams per day—double the highest dose I had ever taken before. I would remain at that dose until I had settled down, and then I would begin tapering down the dose. He was not sure how long I would stay in the hospital.

Then he said something that surprised me.

"We will probably take you off the Imuran, since that could be causing the vomiting."

"I thought the flu was causing the vomiting," I said.

"It may be. This would be more of a precautionary measure. I'll check with your GI before I order any changes, but I suspect that's what he'll tell me."

I was upset by this. I had been taking the Imuran for five weeks or so, and had been initially told by my gastroenterologist—and had confirmed this with my own research—that it could take as long as four to twelve weeks to begin having any effect on the body. If I stopped it now, did that mean I would have to start the cycle all over again, and potentially wait another twelve weeks before I could really begin to see its positive effects?

I kept my mouth shut for the moment, figuring that I would

fight that battle if and when they actually stopped giving me the Imuran.

My gastroenterologist, by an unlucky coincidence, had been on vacation in Hawaii for three weeks when this began, so he didn't visit me until my fifth day in the hospital. Instead, colleagues from his group practice came and saw me to monitor my condition.

On the same day I had spoken with the first doctor, one of his colleagues came in with two residents and gave me a brief examination. He was a young and handsome doctor, the sort you might expect to see on a television soap opera, and his residents were even younger.

The trio sat on the bed opposite me and the doctor began to ask me questions.

"How long has this been going on?" he said.

I thought I could hear an implied criticism in there—how could anyone possibly let themselves get this out of control?—but I could have been imagining it. I reacted to it anyway.

"A long time," I said. "I guess I knew I should have done something, but I was waiting for the Imuran to kick in, and I knew it could take four to twelve weeks, and it's been five weeks now. I was hoping that would begin to settle me down."

"Imuran," he said to me, without even looking up from my chart, "can take three to six *months* to begin working."

Upon hearing that, my heart just sank. I could have another five months of this? I was also embarrassed—he said it in a way that implied anyone should have known *that*. I thought I saw the two residents glance at one another, a silent smirk—stupid patients.

"Really?"

"Yes. You can't really wait, at this stage, for Imuran to begin solving the problem. In fact, we need to evaluate whether Imuran is causing your current problems with nausea."

"I thought that was from the flu."

"Well," he said, "that's possible—but that's why we're here, so we can tell you for certain. I'll talk to your regular GI when he returns tomorrow."

The three of them left abruptly. The exchange had lasted no more than three minutes, and I had never felt stupider and more doubtful of my ability to gain credible knowledge about my medical treatment.

For the next two days I continued to receive the Imuran, so I assumed that both physicians were waiting to let my regular gastroenterologist make the final decision.

But two days later a third gastroenterologist, another colleague of my regular physician, came to see me. I learned from him that my regular physician was back in the office and would visit the next day.

In the meantime, this third doctor's recommendation was that I receive injections of a growth hormone that had proved useful in controlling severe inflammations of the disease. Those injections, a relatively new treatment, would perhaps see me through to when the Imuran would begin to take effect. He would not order them for me, but would recommend them to my physician.

"And will I keep taking the Imuran?" I said to him as he was leaving.

"Of course," he said.

Finally, my own gastrotenterologist returned and came to see me on the day before I left. We exchanged greetings and he asked me some questions about my bowel habits in the last few days.

"I must admit," he said to me after these exchanges, "that I was a bit disconcerted to hear that you were here when I got back. I was expecting things to improve for you after we had started you on the Imuran."

"Me too."

"I found that a little disconcerting," he repeated. He seemed disconcerted. "But it looks to me like you are beginning to heal. I suspect we will let you go home tomorrow, and you can continue the steroid treatments until you are more fully under control."

"Will I keep taking the Imuran?"

"Yes. We'll continue with all of your current medications."

"One of the doctors who saw me recommended some kind of hormone injections. Are we going to do anything like that?"

"I don't think that will be necessary, as long as you are showing signs of improvement."

We spoke for a bit longer about my treatment, he instructed me to set up a follow-up appointment, and we said good-bye.

In the six days I had spent in the hospital, I had seen four doctors and been recommended three different courses of treatment. I had learned my lesson about doctors and medicine in September, in my dealings with the university physicians, but this incident reminded me that such lessons cannot be packed away and confined to my dealings with any particular physician. The uncertainties of diagnosis and therapy will follow me through this disease, and in those areas—and possibly only those areas—however close I am to the bottom, I can never let go.

HAVING TOLD A NUMBER of stories about my interactions with physicians that cast them in a fairly unforgiving light, I must finish this chapter with the story of the one physician who deserves to be named in this narrative, and who has helped to maintain my faith in the medical establishment.

To do that, I need to circle back to the beginning of this chapter.

After the two consecutive nights upon which I had been admitted to the emergency room, only to return home the following day, I received a telephone call from my internist, Dr. Robert Honig.

I had always been happy with Dr. Honig as a physician, and enjoyed my visits to him. He was never a minute late for an appointment, despite his evident interest in the details of our lives—he was always interested in discussing the academic life with me, and always wanted to know about whatever book I had brought in with me to read—and despite his willingness to pay serious attention to whatever medical complaint one brought to him. He explained any treatment he ordered in comprehensive detail, but in terms I could understand. He was the internist for a number of us at the college, and all of us had stories testifying to his unusually personable and attentive treatment of his patients.

"How you doing, Professor?" he asked me, when he called me that morning.

I explained what I had been going through the last couple of days, and we agreed that I would let him know if I saw no signs of improvement within the next day or two, or any changes for the worse.

The next morning my mother, my wife, and I all decided that we could not continue dealing with this level of disease activity on our own. We called Dr. Honig and let him know we thought it was time for me to return to the hospital.

It was a Saturday morning. I telephoned the call line—his office was closed on Saturdays—and he returned my call within ten minutes.

"Go to the hospital now," he said. "I'll meet you in the emergency room. We'll get you in faster that way."

In the emergency room a nurse checked my insurance information, and then instructed me to wait until an admitting desk was free. We sat down; a few minutes later, Dr. Honig came in and greeted us, shaking my hand as he always did.

"Let me see if I can speed this process up," he said.

He went over to the admitting desks and spoke to someone there for a moment. Then he waved us over, and stood behind us while they took additional information from us. The admitting clerk told us that I would still need to wait for a hospital bed, but that they would get me into one as quickly as possible.

Dr. Honig disappeared behind the administrative doors of the emergency room, and was back shortly to let me know that some beds were opening up, and I would be taken upstairs soon. Sure enough, ten minutes later I was on a mobile bed, IV in my arm, waiting for an orderly to transport me upstairs.

Dr. Honig promised to return in a few days.

On my third day there he walked through the door of my room late one afternoon with a smile on his face and his usual greeting:

"How ya' doing, Professor?"

"I'm getting there."

"It sucks, doesn't it?"

He sat down in the chair next to my bed and spent a few minutes commiserating with me. He wanted to know how I was handling the time I was missing at school, and how my wife and children were holding up. I was half waiting for him to begin his examination, and give me some news about my treatment, but he never did either of those things.

After he left, I realized he had never examined me. He had spoken to me about my treatment only because he seemed curious to know what I had been told by the attending physicians and gastroenterologists. The moment I understood this was the moment I realized that he had been visiting me only in part as my physician—he had also been coming to visit me as a friend. I was a human being to him, not a disease process.

As I reflected afterward upon his treatment of me throughout this period, it occurred to me that he, too, had been letting go to a certain extent—to the same extent, actually, that I had needed to let go.

He did not abdicate his care of me. He confirmed my decision to enter the hospital, he saw me through the doors of the emergency room and into a hospital bed. And I have no doubt that he checked my charts to monitor my treatment.

But he recognized, too, that my disease had proceeded beyond the point at which he—as an internist—could help me medically, and that my treatment belonged in the hands of the specialists who had greater familiarity with my disease.

From him I learned the value of the physician who knows how to let go. I have dealt with at least four different internists since I have had my disease, and have watched them attempt to assert different levels of control over my medical care. While I don't want an internist who sends me running to a specialist for every twinge and stab of pain I feel, I do want someone who recognizes when my care needs to move to the next level of specialization.

Observing his care of me, and seeing his actions and attitude parallel the one I was developing during that period, reaffirmed for me what I learned over the two weeks I spent in and out of the hospital: When you are at the bottom of the well, you need someone who knows the contours of that well to throw you a rope. It remains my responsibility to grab the rope when it comes, but I will always be the first to know I am down there.

I need to send up that cry for help, and to trust that the rope will come.

7

THE IMAGINATION
OF DISEASE

Learning to Be a Father

[February–April 2001]

One unfortunate consequence of my hospitalization was that it brought a temporary halt to our efforts to conceive a third child. Anne and I had been making halfhearted efforts toward that goal until that point. The "half" part of halfhearted came from me—Anne was far more enthusiastic about the possibility of this third child than I was.

I felt I had good reason to be halfhearted. Crohn's disease has a clear genetic component; 20 percent of sufferers have a relative who has the disease as well. The disease did not manifest itself in me until I was in my early twenties, and in that respect I was lucky. It can strike children as young as three or four years old. For those unfortunates, childhood can be a painful experience, physically and emotionally. Place yourself in the skin of a child who has to rush from the classroom frequently for trips to the restroom, who might miss weeks or months of school to illness, who might suffer excessive acne, facial bloating, and even stunted growth as a result of his or her medications.

As any parent will easily imagine, one of the most emotionally disturbing visions of my future now centers upon the prospect of seeing my children suffer from this disease. To envision any of my children undergoing what I have experienced this past year is so acutely painful to me that I wince at the thought of it.

As much as she empathizes with my condition, Anne simply will never be able to understand this fully. She has not experienced the disease as I have, and she will not be the one whose faulty genetic code may lead one of our children to a life of disease. Of course it would be irrational to see myself as somehow blameworthy if one of my children should inherit the disease, but I am afraid those feelings would overwhelm me all the same.

So while part of me shared Anne's desire for a third child, part of me resisted. My stay in the hospital strengthened the case for both of these parts: Seeing the bottom of the well increased my anxiety about the possibility of a child suffering that same fate; spending time away from my children helped me remember how much they meant to me, and sparked a desire to continue to see our family grow.

In the weeks immediately following my stay in the hospital, I felt no pressure to make any concrete decisions about whether or not we should redouble our efforts to conceive another child. I barely had enough energy to get my body up the stairs, much less spend time in bed with Anne.

I had another, far more immediate concern to worry about in those two weeks: the two children we already had.

FOR GOOD OR ILL, a combination of circumstances prevented me from understanding the effects of my hospitalization—and my illness in general—upon Katie and Madeleine until I had been home from the hospital for at least a week. When that understanding finally came, it was as devastating to my emotional stability as the illness had been to my physical health.

While I had been in the hospital, Anne had continued to teach full time, and, with the help of my mother, to pack Katie off to kindergarten and Madeleine to daycare every day by herself—not to mention cooking, taking care of the house, and worrying about me.

We had kept the girls away from the hospital because of the lack of entertainment for them there, and I suppose from an unstated agreement that they were not yet ready to face this part of my life.

When I did return home from the hospital, on a Thursday morning in late February, I was extremely weak and fatigued, and had trouble making it up and down the stairs more than a few times a day. Lying in a hospital bed for six days, along with the effects of the disease, had robbed me of most of my strength and energy. I had four days to recover before I had to be back in the classroom for one week until the semester's spring break. When spring break came I would be able to rest and recover at a more leisurely pace.

Hence I was not much of a playmate for the girls for those first few days, and beyond that for the first few weeks, after I returned home. While they began to venture outside to play, in the slowly warming weather of a New England spring, I couldn't do much more than pull up a chair on the back deck, huddled in my jacket, and watch them. Moreover, I had fallen so far behind at school that I had to reserve any energy I could muster for catching up on my grading and class preparations.

Through all of this, I had been happy to see that neither of the children ever seemed very strongly affected by my illness. When they visited me that one time in the hospital and when I watched them from the couch in our family room, they continued to do the sorts of things that small children always do: play with one another, play by themselves, fight over toys, demand snacks and juice, and pop over to give me the occasional hug and kiss. But they did not seem to me especially concerned about what was happening to me—or at least they never expressed such concerns to me or Anne.

I was quite content with this. I did not need my children feeling sorry for me and worrying about me—I had enough family members and friends already doing that. The last thing I wanted at that time was the additional burden of thinking that I was causing them psychological distress. I was happy to see them continuing with their lives as if nothing unusual were happening.

I have learned throughout my eight years or so as a parent that children can be extremely perceptive, and that even the smallest anxieties can weigh heavily on their minds. I am not sure, then, why I allowed myself to be so blind to how my hospitalization was affecting my children, my older daughter in particular. I suppose perhaps I needed that blindness; at that time I had just enough psychological energy to maintain my own balance and had none to spare for the rest of my family.

I had one early warning sign of what was happening to Katie just a few days after I returned home from the hospital. On Monday she came home with a picture from school that depicted my return from the hospital, with our family all together again. I asked her why she drew that particular scene, and she responded:

"We were supposed to draw the happiest day we could remember."

That drawing made me suddenly aware that, contrary to what

I had imagined, Katie had really been aware of my absence from home. Neither my wife nor my mother had reported to me that the girls had been excessively mournful during that absence, so I had assumed that—while I knew they loved me and missed me— they had been able to fill that gap with the excitement of having their grandmother in town.

And that may have been partially true. But my absence had made more of an impact on Katie than I had suspected.

The full extent of that impact did not become clear until several weeks later, when my wife ran into Katie's teacher at a conference. She learned then that, during the week I spent in the hospital, Katie had been upset and anxious at school, complaining of stomach pains, and had even cried on one occasion—something she rarely did in class.

The Monday after I came home from the hospital, she complained to her teacher that she had a stomachache, and asked if she could be sent home. Her teacher knew that I was recuperating from an illness, and decided that it would be better if she could find a way to keep her at school. She pressed her about how she felt, at which point Katie admitted that she did not have a stomachache at all, but wanted to be sent home so she could spend time with me.

"I miss my daddy," she told her teacher, beginning to cry.

When my wife reported this to me—and I could not hold back a few tears myself—it was like hearing a piece of bad news which, somewhere deep down, you knew was coming.

Over the next several days, I tried to involve Katie in a conversation about how she had felt when I was away, in the hope that I could perhaps relieve any lingering anxieties. I repeated to her what her teacher had told me, and asked her if it were true that she had been upset and crying at school.

"Yes," she said simply. "I missed you."

But she wouldn't elaborate any more than that, despite her normal tendency to chatter on endlessly about everything in the world. I suspected that forcing her to remember and think about the whole incident was causing her additional distress, so I stopped speaking with her about it—though that didn't stop me, of course, from worrying about it.

MADELEINE'S RESPONSE to this situation, in the meantime, could not have been more different; her contrasting reaction undoubtedly stemmed from her own particular stages of emotional and intellectual development.

Madeleine has always preferred her mother to me. My guess is that this is because Anne spent more time at home with Madeleine after her birth—over four months—than she did with Katie. Moreover, I spent one or two days a week at home with Katie when she was an infant; after Madeleine's birth, I was working full time. These are my speculations, but who really knows?

My stay in the hospital entrenched Madeleine deeper in her patterns of affection. When I finally began to spend time with the children again, it was apparent that Madeleine had grown unused to me.

In my interactions with her, I felt more like the uncle who occasionally visits from afar than the father who lives with her: the sort of adult figure who inspires a distant—because obliged— affection, some fear, and uncertainty.

Consider a typical morning, in late March, two weeks after I had returned to school from my spring break, and nearly five weeks after my stay in the hospital. As usual, Anne awakes before I do to prepare herself for the day. The girls wake up when she does, and so she sends them downstairs to watch television until one of us comes down to make their breakfast.

I pull myself groggily out of bed when Anne begins drying her hair, at which point sleep—thanks to the noise of the hair dryer—is no longer possible. I stumble downstairs and, over the protests of the girls, switch off the television in the family room.

"You can watch it while you eat your breakfast," I say. "Come into the kitchen."

I turn on the small television on the kitchen counter—the girls settle immediately into their chairs at the kitchen table, mesmerized again by the glowing screen—and pull together some breakfast foods: a bowl of Cheerios and a half of a banana for each of them.

Katie begins to eat as I set the food in front of her, but I have to tap Madeleine on the head and point out that her breakfast awaits. She turns her full attention away from the television and looks at the breakfast I have made. Her mouth sets.

"I don't want that," she says quietly.

"Madeleine," I sigh, hopeful that I can reason with her, "you *like* Cheerios and bananas. What's wrong with this breakfast?"

"I don't want that," she repeats.

I react quickly. Stupidly.

"Fine," I say, "then you don't get anything. You can starve."

She won't look at me.

"I want Mama to make it."

I knew that was coming, and it frustrates me. Of course I should react with patience and understanding—but it's 7:30 AM and I don't have the time for this and I want to be helpful to my wife, to let her experience a few moments this morning when she will not have to make breakfasts or dress her children, when she might be able to sit and have some coffee and some cereal. For the past six weeks I have been almost completely useless around the house, and now that my strength and energy are returning I want to begin to carry my load again. Madeleine is not letting me do that.

"Mama is busy; she's drying her hair. Mama doesn't have time to do everything for you. *I* can make your breakfast. *I* can help you out. *I* love you, too, you know."

My words come out angrily, not how I want them to sound, even though I am indeed feeling anger. I want to sound patient, reasonable, understanding. Katie picks up the tone in my voice, and she is watching now, paying attention to my words and to Madeleine's reaction.

Madeleine's lower lip begins to quiver. "I want Mama to make it."

I can hear the hair dryer stop upstairs, which means that Anne will be down in a minute or two. Do I fight this battle?

"She always has to have Mama, Dad," Katie says to me sympathetically.

And if I always let her have Mama, will she ever change? Do I force her to let me help her, giving her no choice in the matter—*eat what I serve or eat nothing*—so that she gets used to the idea? Or will the use of such ultimatums simply turn her against me even further? Do I simply wait for her to come to me for help on her own?

But what if she never comes?

I am standing at the counter contemplating these questions, watching Madeleine stare steadfastly into her unwanted bowl of Cheerios, unwilling even to look at me. I can hear Anne coming down the stairs, and I am still meditating on how to react when nature intervenes with its usual demand and I step quickly into the bathroom. By the time I get out Madeleine is munching happily on a piece of toast that Anne has made for her. The half of a banana and full bowl of Cheerios are sitting, untouched, on the counter.

When the time comes to get them dressed, we don't have the luxury of pondering the subtleties of the appropriate parenting strategy: I dress Katie, and Anne dresses Madeleine.

Anne is leaving before me this morning, so I help gather together backpacks and lunches and bundle everyone toward the door. I snatch a quick hug from Katie, and plant myself in front of Madeleine as she scurries to follow Anne out the door, trying to attach herself to Anne's pant leg.

"Can I have a hug?" I say.

She pushes her body around my leg, trying to worm her way out. I reach down and give her a crushing bear hug. "I love you anyway, Madeleine," I say, and kiss her on the cheek.

She smiles just a little bit—just enough so that I know I still have a chance—and then she is out the door and down the walk, chasing after Anne, calling "Mama! Mama!"

This behavior continues, it seems to me, for weeks and weeks. And for weeks and weeks I am frustrated, angry, depressed, and baffled by it. I don't really know how to react, and Anne doesn't either.

I am saved, in the end, by what Madeleine calls our "projecks."

IT IS A LATE APRIL DAY, a Wednesday afternoon, and the weather seems like it is just about ready to begin behaving as if it were spring. Green buds and shoots are making their appearance in the bushes and trees around the neighborhood, and the temperatures have been climbing consistently enough that we have confidently put away our winter jackets.

I have picked up Katie from school at 2:25. Together we drive to pick up Madeleine from her in-home daycare provider, and we are home by 2:45. Katie wants desperately to play with a neighbor girl her age, so I arrange that quickly and Madeleine and I walk her down the street to her friend's house.

Back at home I am tempted to turn on the television and lie on the couch—it has been a long semester, and I am often still tired—but for the past few weeks I have had something in mind for Madeleine, and this seems like the most opportune time.

We had brought back a box of shells from our recent vacation in Florida at my parents' condominium two weeks ago, and I have been promising her that we will do one of our projects with the shells.

"Get your box of shells, Madeleine," I tell her.

"Are we going to do a projeck, Dad?" she asks me. Her eyes are wide with anticipation.

"Get the shells, kiddo."

She goes dashing off to find the shells, and I find the large poster board I have bought in anticipation of this event. I spread it out on the dining room table with some construction paper, a tube of glue, scissors, and a box of crayons. I begin cutting out fish shapes from the construction paper, drawing little smiles and eyes on them in profile.

Madeleine returns with the box of shells, and we settle down side by side at the table.

"What are you cutting, Dad? What are *those*, Dad? Are they *fish*? Can I *help* you do that, Dad?" She says it with such a sense of enthusiasm and anticipation that I can't help but smile.

"We're going to make a beach project today, Madeleine. Let's color in the sand and the water, and then we'll glue on the shells and the fish."

"I can do it! I can do it!"

She is almost desperate with anticipation, bouncing up and down in her seat. She doesn't even know what it is she is claiming she can do; she just knows that whatever we're doing, she wants to be involved.

"Okay, kiddo," I say. "Settle down. We'll do it together."

I draw a beige line across the middle of the poster, and then give her the crayon.

"Color in the sand, all below this line," I tell her.

She begins scribbling furiously, covering the poster in long strokes of sandy crayon. I have to keep pointing out the places she has left white. We repeat the same process for the water and the sky. As she colors I continue to cut out more shapes: an octopus, a dolphin, a starfish, a sea turtle.

Finally we arrive at the moment she has been waiting for: the gluing. She is obsessed with gluing things together.

"Not too much glue now," I tell her, though I know the warning is futile.

She holds the tube of glue over the back side of a fish, a look of determination in her eyes, her little hands squeezing with all their might. The glue comes squirting out, too much; at just under three, she does not yet have the rational capacity to understand that she must stop squeezing *before* the required amount of glue actually hits the paper. The glue coats the entire back of the fish, thickly.

"Madeleine!" I say, too loud, wiping off the excess with a paper towel. "Just a little bit of glue!"

"Sorry, Daddy," but not really meaning it, breathless with anticipation at the prospect of gluing another fish. I have to remind her that, once she has applied the glue, she actually needs to paste the fish into our colored ocean.

And so we work through the project, slowly, Madeleine gluing everything she can get her hands on, me occasionally losing my temper at the mess she creates but mostly enjoying myself. For the shells I fill a Dixie cup with glue and we use toothpicks to dab it along the edges of the shells. We scatter them around the beach, scallop and mussel and clam shells and sand dollars. When we are finished I take a brown marker and carefully etch into one corner our signature project line: "Madeleine and Daddy 4/19/01."

She is immensely proud of it and wants to hang it immediately on the wall downstairs, where the children's artworks provide the decorations for the finished portion of our basement. I have to convince her to wait, explaining to her for the hundredth time how glue has to dry.

Over the next few months Madeleine and I will do project after project, most of them involving gluing something onto something else: I draw a picture of a tree, we collect leaves and acorns and glue them onto the picture; I cut out the shape of a butterfly and we glue scraps of paper to its wings; I collect shiny objects and we glue them to paper plates to make alien spacecrafts.

And over the next few months, I gradually work my way back into Madeleine's affection, gluing together a relationship that I once feared had been irremediably splintered by my illness. The rebuilding of my relationship with her reminded me so much of gluing that it was uncanny—only this time, I was the one who needed help understanding how to apply the glue properly.

In the mending of our love, I wanted to take the glue and dump it all in at once, hoping that I could resolve everything quickly. Madeleine forced me to glue properly: to apply the glue slowly, in small dabs, and to be patient and wait for the first drops to dry before applying the next ones. With each project I dabbed another drop into the crack, and with each of those drops the crack sealed a little further, and held a little more firmly.

Madeleine and I are friends now. She still has a special relationship with Anne, and no doubt she always will, but she and I have our own special kind of relationship. I am back in her life, and—at least for the moment—she believes that I'm here to stay.

THE FIX FOR KATIE was nowhere near as simple as this. The difference between two or three years old and five years old is significant, and I couldn't paper over the memories of my hospital stay in her mind with some glue and construction paper. At the moment I am writing this sentence, over a year after hitting bottom, I am quite certain that Madeleine—now four—has no recollection of my stay in the hospital; I am equally certain that Katie does.

Initially, all I could provide for Katie was my presence, both emotional and physical. I simply had to be there with her as much as possible in order to dispel the anxiety that I would go away again. I understood how important that presence was to her one afternoon in the late spring when I had picked her up from school.

At her school the parents wait for their children outside a chain-link fence that encloses the playground, and a couple dozen of us—mostly the parents of kindergartners and first- or second-graders—gather there every afternoon to wait for the bells that send our children out to us.

Many of the parents, almost all mothers, know one another. As I stand at the fence I listen to them talk about their children, about neighborhood gossip, or about the weather. I mostly keep to myself; I know one or two of them, but I am content to stand alone and watch for the kids to spill from the front doors, laughter and bundled energy bursting in little streams from the school.

On this particular afternoon, a happily warm and sunny one, Katie spies me immediately and smiles, walking quickly over to where I stand waiting. As soon as she reaches me outside the fence, she slips her hand in mine and pulls me down the sidewalk toward home. For some reason I take note of her action, and become especially conscious of the warmth of her small hand in mine.

It's a ten-minute walk home, and holding hands with a smallish

five-year old, when you are weaving your way through hordes of
children and teenagers coming from school, can be a difficult trick.
About halfway home, caught in the middle of a group of jostling
middle-schoolers, I drop her hand.

For a minute or so, we continue whatever conversation we are
having without interruption. And then, without a word or a pause
in her step, she reaches up again and grabs my hand. We hold
hands the rest of the way home.

That second grab of my hand catches my attention, and I begin
to notice that, in the weeks and months following my hospitalization, holding hands becomes a habit for us. It clearly offers her a
reassurance that she needs, and I am glad to make it available to
her.

But, of course, I won't be around to hold her hand all the time.
At some much later point, too, I know she won't want to hold my
hand anymore. The reassurance of my presence provides her with
some relief, but I can see clearly that it is a temporary solution.

It is a temporary solution for me as well.

Chronic disease induces hypochondria not only for my own
health, but also for the health of my children. All parents worry
about their children's welfare, and I am no different. But I have
the added concern—as do many parents with genetic disorders or
diseases of any sort—that my children are especially vulnerable to
my specific condition.

Every time a child complains to me of a stomachache, or diarrhea, or excessive gas, or of an urgent need to use the restroom,
my radar goes up. Until I see a complete return to normalcy, that
complaint lodges itself into my brain and tugs at the strings of my
paranoia and hypochondria. Once, on vacation in St. Louis, I
thought I saw a reddish streak in Madeleine's bowel movement;
when we returned home I immediately took her to the doctor, who

could find nothing wrong. When Katie was four years old she spent the spring complaining of stomachaches. Anne assured me that this was the commonest of childhood complaints, and likely meant nothing. I wasn't satisfied; I took her to the doctor, who told me that stomachaches were the commonest of childhood complaints, and that he could find nothing wrong with her. Eventually she stopped mentioning them.

But those hours and days and weeks of worry can't be so easily dispelled, even by a happy diagnosis from a pediatrician. At night I stand in their rooms and watch them sleep, their covers thrown off, arms and legs splayed across their sheets and blankets. Sometimes I raise my hand and make a sign of the cross in the air, a layman's blessing—first Katie, in the top bunk, then Madeleine below. Mostly I just stand and watch, knowing, too, that the days and years in which I have this privilege of watching my children sleep in perfect health may be numbered.

What disturbs my vision, especially after my time in the hospital, is imagining the children living through a similar ordeal, for any reason, or having to spend weeks or months laid up in bed. What a change from their lives of running and dancing and laughing it would be! How could they survive it? How could I survive it?

It was not until later in the year, during the driving vacation we took to visit our old home cities in the Midwest, that a kind of solution to both sides of this problem began to present itself to me. A solution of sorts—nothing, I understand, will ever remove my fears for my children, or, as they get older, their fears for me.

But I learned something, on that trip, that might help.

OVER THE COURSE of that trip we spent, in transit between Massachusetts and Cleveland, St. Louis, and Chicago, somewhere in the range of forty hours in the car. We did everything we could to make those hours more palatable for the girls. We bought them small toys to play with in the car; we carted a case of their dolls along; we packed their favorite CDs; we checked out from the library some children's books on tape; and we brought a shopping bag full of videos for the TV/VCR that came installed with our minivan. We tried always to make sure we were on the road during the hours after lunch, when body metabolism slows and even Katie is likely to take a nap if she's immobilized in her car seat.

But even with these precautions, it was impossible to banish all dead time from the car: those moments when the videos are giving them motion sickness, no one could stand another second of the soundtrack from *The Little Mermaid*, their dolls are scattered on the floor around them, and naptime is over.

So in those moments we did the only thing we could do: we stared out the windows. And, especially as we drove through the mountainous and tree-lined highways of Massachusetts and upstate New York, a memory of a childhood activity began to emerge from the fog of memories I have of my youth.

For many years the family vacations we took when I was a child were brief trips to a small lakeside cabin within a couple of hours from our home, and so did not involve excessive amounts of driving. But one year we drove from Cleveland to Florida over the course of three days, and for another few years we went to an island off the coast of South Carolina. We did not have the luxury of VCRs or even tape decks during those drives, and a propensity for motion sickness in my family prevented almost all of us from reading. So we fought with one another, and played travel games, and slept.

But as I drove with my own children through the window-staring hours, I gradually remembered how I had occupied those dead times when I was young, in an activity that fell somewhere between a game and a daydream.

I see very clearly now that this activity was the mental calisthenics of a budding writer. As we would drive through any particular landscape, I would picture myself suddenly set down in the midst of it, and imagine what might happen to me. Driving between the dynamite-blasted walls of a mountain, I would watch myself meticulously climbing the rocks, away from the speeding cars to the safety of the woods. In the fields of Midwestern farms, I would see myself threading my way through a herd of cows, in search of a farmer who could help me find my way back to my family. In the cities I saw myself set down at a busy intersection, alone and lost among the tall buildings and crush of horn-blowing cars. Dropped into the midst of a forest in a valley, I would find a running stream of water and follow it to civilization.

The nature of my imaginative flights from the car eventually evolved from these hypothetical survival narratives to the more enjoyable pastime of simply envisioning myself as a resident of any particular landscape. Now I was a farmer's son, rising every morning at the crack of dawn to milk the cows and collect the eggs from the henhouse; now I was a hiker alone on the mountain, cooking my food over an open fire and sleeping under the stars; now I was a boy who lived in an apartment in a city, and saw nothing but concrete and steel everywhere I turned.

At one point in this most recent vacation, I realized that these memories were returning to me because, in fact, I have never given up this imaginative exercise. I still play these games in my mind on long car trips. Before we set out for such a trip I usually will identify some intellectual problem I need to think through, so that I can use the time in the car productively—do some advance planning for a syllabus, for example, or decide upon the right ending for a piece of writing I am finishing. Instead of thinking diligently about these problems, my mind inevitably wanders away, seeking out shelters in the landscapes around me.

As we drove back to Massachussetts from Cleveland, I introduced my daughters, especially Katie, to this speculative exercise. At the time, I did not do it for any conscious purpose other than to help us pass the time in the car. It was easy enough to play this game as we drove through the mountainous terrain of upstate New York and western Massachusetts, passing by one spectacular vista after another along the Massachusetts Turnpike: imagine yourself at the top of that mountain, imagine yourself at the bottom of that one, imagine yourself on a boat in that river, imagine yourself living in that lonely farmhouse on the hillside there.

As we imagined our way through those landscapes, a series of linked ideas began to emerge and crystallize from the reflections on fatherhood I had been entertaining since my hospitalization. I started to see a link between what had helped me pull Madeleine back into my life, what I was teaching Katie on those car trips, and what I could offer to my children to help them cope both with my own illness and with—God forbid—any future illnesses of their own.

As a writer, I have probably always been aware that the greatest gift I can offer to my children is the imagination, and a full appreciation of the satisfaction that comes from artistic creation of any sort. That realization only came fully to my consciousness, though, as the long shadow cast by my hospitalization began to recede.

Now I see clearly how I can best be a father to the children of a chronically diseased man, one whose genes may one day put them at risk for that same chronic disease. I cope with my illness through my writing, as I describe more fully in the next chapter, and through the process of creating artistic order from the senselessness of disease. Writing helps me both work through the disease and escape from the disease.

I want my children, whether the time ever comes that they

need those skills or not, to be equipped with the imaginative and creative powers to help them cope with whatever trauma they may face in their lives. Gluing acorns on a badly drawn construction-paper tree models those creative powers at the most basic level, as does cultivating the ability to cast one's imagination out of a confined and closed space into the landscapes you will never touch, and perhaps never see again.

Before my hospitalization, I guided my children through those activities on instinct, from the common desire that all parents have to lead our children along the happier paths that we have followed; I do so deliberately now, nurturing their creative and imaginative powers with projects, with shared efforts at writing stories and poems and songs, with oral rhyming games and other word play, and with, above all, a constant modeling of my own love for books and the works of the creative imagination in any form.

Of course I want my children to take a healthy interest in the outdoors, in exercising their bodies, and in sports and normal childhood games. But most of all I want them to have the creative and imaginative resources to thrive and enjoy their lives, whatever fates their bodies hand to them.

As THIS LESSON became clear to me in the spring and early summer, I was able to soothe my anxieties about the genetic inheritance I might pass along to my children, and to turn with a willing heart to the task of trying to conceive our third child.

But I want to save that story, and its outcome, for the final chapter.

8

CARRYING OUR SECRET PAINS

Learning to Tell My Story

[April–May 2001]

Around eight years ago my mother was diagnosed with cancer in her tongue and her lymph nodes, and went through a successful course of surgery and radiation therapy to remove the growths. Still, though, she occasionally gets small growths beneath her tongue that require surgical removal; those removals have ranged from small outpatient procedures to skin grafts requiring reconstructive surgery. Her doctors have told her that she may continue to see these small growths for the rest of her life, but that they are treatable and should never become life-threatening, as long as they are closely monitored.

Maddeningly—at least to me—she often does not tell her children about these growths, and sometimes even the minor outpatient surgeries to remove them, until after they happen. Then she glosses over them quickly, letting us know via e-mail that she has seen the doctor and that everything turned out well.

"Mom," I will say to her afterward, frustrated, "I ask you every week how you are doing, and you never said anything about this new problem during the entire month you were worrying about it. Don't you think I really want to know how you're doing?"

"I know," she will respond simply. "I'm sorry."

During one of those discussions, I managed to squeeze a little more out of her about why she does this.

"I know you don't feel this way, Jimmy," she said to me over the phone one evening. "I know you understand what it's like. But most people don't want to hear about it when you're sick, or when you're worried about being sick."

"I know," I said.

"So when people ask me how I'm doing, I just tell them I'm doing fine. And that's what they want to hear. They want to hear I'm doing fine. People don't want to listen to me describing the little bump I discovered under my tongue last week, and hear me complain that I'm not sleeping at night because I'm so worried about it."

"You're right, Mom," I said. "I've had the same experiences. Just remember that I'm the exception. I *do* want to hear about it."

"I'll try to remember," she said.

As I reflected upon this conversation later that afternoon, it took me just a moment or two of self-analysis to realize that I was often guilty of the precise sin for which I was chastising my mother.

MY SISTER GOT MARRIED in October of 2000—around the time described in chapter four—the last sibling of the five of us to do so. We flew to Cleveland for the wedding, and at the reception

I had a chance to speak with members of my extended family whom I had not seen for many years.

One of those was my mother's brother, Uncle Mike, whose wife had recently died of cancer. I stopped to talk with him on my way to the bar for a drink, and he asked about my health. At that time I was taking quite a bit of medication, unsuccessfully trying to taper off prednisone and dealing with moderate disease activity, and I told him so. My mother had shared with him the details of my illness, so I probably offered him more detail than I would have to most friends or acquaintances who ask me about my health.

After hearing my health report, he made a sympathetic comment about the hassle of taking multiple daily medications. I responded with some version of the comment with which I almost always conclude narratives of my health and my medications:

"It was hard at first, but after a while you get used to it. It's not so bad anymore."

He smiled at me, and nodded.

"I'm sure that's what you tell people, Jim, but I know that's not really true."

I was momentarily taken aback at his response, but then recollected that he had spent the last year or two caring for his dying wife.

"You're right," I said. "But that's what I tell people."

I HAD BEEN TELLING PEOPLE that for five years, for several reasons.

Like most young people—even those of us as "young" as thirty-three—I had youth's fear and revulsion of infirmity and disease. Prior to my diagnosis, the malfunctioning and decay of the human body in any form repulsed me, no doubt because it reminded me of the fact that I, too, had a physical body that would one day suffer this fate.

I knew that other people shared these sentiments. I had seen them in the way that children, and even college students, react to physical decay and disease. Students in my writing classes will often write essays about their first experiences with death, as they have watched their grandparents die. Those essays frequently describe the mixed feelings students have at reconciling their love for their grandparents with their revulsion at dealing with a decayed and diseased human body.

I have a nightmarish series of memories of this sort: a collection of mental slides of the death of my grandfather, including—hitting far too close to home—a scene in which he has shat himself, and is calling both plaintively and angrily to my grandmother to come and change him; another of him, frail and emaciated, in a nursing home bed, unable to remember or recognize me—me, who had for many years privately bestowed upon him that important childhood honor of favorite grandparent.

So my primary fear was that any revelations I made about my condition would cause others to see me, first and foremost, as a diseased body. I knew what could accompany that perception: the unpleasant reflections on the parts of our physical selves that we do our best to ignore, and the fear of our own mortality. I had felt those same emotions myself, in the face of others' pains and illnesses.

But I was also afraid that even those people who might not find the diseased body repulsive might still brand me with that label, and that prospect was equally unpleasant to me. I was afraid that, for most people, I would be lodged in their minds as "the guy with Crohn's disease."

That label captured the one part of me that I wanted to dissociate myself from. Label me the Professor, the Father-with-Two-Children, the Writer, the Brother and Son, label me Quick-Tempered, Lazy-about-Housework, Struggles-with-Alcohol, Occasionally Self-Centered—label me as anything but Man-with-Crohn's-Disease. No single label could encapsulate what seemed most essential to my self, but the disease label seemed to me especially distant from the parts of myself in which I could take some pride and satisfaction.

So, for five years I had been doing my best to conceal the actual nature and extent of my illness from both friends and colleagues at school. Of course my family knew, and some of our closest friends, but I tried to minimize the number of people who were fully aware of my condition. When I was ill I relied on the stock phrase "I wasn't feeling well"; if I needed to explain in more detail a prolonged bout of sickness, I simply said that I had "stomach problems."

Even then, I knew I was doing precisely what so angered me when I read about others coping with the disease—papering over my illness and its symptoms with easy euphemisms. But that didn't stop me from turning around and using those euphemisms when I was confronted with an acquaintance curious about my health.

I maintained this strategy of denial and evasion even as late as my stay in the hospital, when—you would think—I would no longer be able to hide the nature of my condition. But remember that e-mail I sent to my colleagues from my hospital bed, asking them to help me with my classes? Here is the paragraph in which I described what had put me in the hospital:

> I am in the hospital at the moment, and will be here at least until Wednesday. I expect to be out on that day, and back in the classroom either on Friday or Monday. FYI, I came down with

an awful stomach virus last Monday, which put me in the hospital on Monday night and Wednesday night—only to come again on the following day each time. Mistake. I am now in here for a little bit of a duration—the stomach virus triggered some other problems, and I am getting IV fluids and medicine for a few days until it has all settled down.

"Some other problems": five years of illness, five years of multiple medications a day, five years of anxiety and concern about my health, five different gastroenterologists, three colonoscopies, one week in the hospital, all rolled into that innocuous little phrase.

"Some other problems." No phrase could capture more perfectly the extent to which I would seek out meaningless phrases to describe my illness to others. Anything that avoided reference to the chronic nature of the problem would do; anything that avoided reference to the specific nature of the disease was absolutely necessary. Stomach viruses were acceptable; inflamed colons were not.

And this leads to the final reason I did my best to conceal the disease from others—the fact that the primary symptom of this disease, at least in my case, is one that we don't discuss in contemporary American society.

Proof of the silence our culture steadfastly maintains about diarrhea is most apparent in television commercials for antidiarrheal medications. Most of the time, those commercials won't even mention the word diarrhea until the final few seconds, when a picture of the medication comes onto the screen—if they even mention it at all. They tend to play out like this:

SCENE: Good-looking man and wife are dressing for a party. The man suddenly sits down on the bed, bow tie hanging over his tuxedo shirt, and says to his wife:
MAN: I don't think I can make it tonight, honey.

WOMAN (making a face): Is it . . . ?
MAN (nodding ruefully): Yes.
An announcer breaks in to explain authoritatively how the medication coats the lining of the stomach and intestines with soothing medicine, providing relief—from what?—for up to twelve hours. Cut back to . . .
SCENE: Man and woman are standing together at a formal party, laughing happily. FADE OUT . . .

The woman's pained expression, her inability to complete the question—to use the actual word—reflects one of our society's great taboos: we do not, in public, discuss our shit.

I have found the best explanation for this social taboo in the work of a now-deceased philosopher and psychologist named Earnest Becker, whose Pulitzer Prize–winning book, *The Denial of Death*, I first encountered as an undergraduate—at precisely the time when I was experiencing my first regular and sustained bouts of diarrhea. Those bouts were nothing like what I experienced in the weeks preceding my diagnosis in 1996, but I suspect they were the first manifestations of the disease. At the time, I attributed the diarrhea to dining hall food and too much beer. But, whatever the cause, the diarrhea I was experiencing at that time no doubt highlighted for me the lessons Becker has to teach about the meaning of shit in the human condition.

Becker argues that human excrement reminds us of two features of the human condition that we all would like to ignore: the limitations of our human aspirations to intellectual and spiritual achievement, and the inevitability of our physical death and decay.

Describing the paradox of humanity succinctly and eloquently, Becker calls the human animal "the god that shits." We are godlike in what we have achieved in spiritual and intellectual matters: the finest works of art and literature, the amazing advancements in science and technology, and the moral heroics of our earthly saints. But however lofty our intellectual and spiritual ambitions may be, however incredible our achievements, we all must acknowledge our identity and limitations as human bodies when we are on the toilet. When Michelangelo descended from the ceiling of the Sistine Chapel to move his bowels, he shared the fate, if not the actual toilet, of the meanest of Italian peasants—not to mention our own fates as well.

Excretion reveals to us too that our free will, as human animals, is limited by our bodies. Although we have basic bodily needs in many areas—such as the needs for food, drink, oxygen, and warmth—those are all needs we can in fact choose to ignore. Ignoring such physical necessities might lead to our own self-destruction, but we are perfectly capable of making that choice. I may refuse food and drink, and I may choose to end my own life by depriving myself of oxygen.

Not so with excretion. Excretion makes its demands upon the body, and requires that we subject ourselves to its whims, at least momentarily. Although we may be able to postpone the body's demand to excrete, we must succumb to it eventually. It is for this reason that Becker places excretion at the focal point of the conflict between our physical bodies and our spiritual and intellectual values:

> Nature's values are bodily values, human values are mental values, and though they take the loftiest flights they are built upon excrement, impossible without it, always brought back to it . . . The anus and its incomprehensible, repulsive product represents not only physical determinism and boundness, but the fate as well of all that is physical: decay and death. (31)

When we excrete, we acknowledge the submission of ourselves to the natural world's cycle of birth, decay, and death. This natural cycle exists and perpetuates itself without our formal consent—it pre-existed us, it controls a significant portion of our daily lives, and it will outlast us. Shitting is the closest and most frequent contact most of us have with the fundamental paradox of being human, caught between the world of physical necessity—and its attendant features of decay and death—and the world of spirituality and intellect.

I am not suggesting, of course, that the advertising executives who manage the antidiarrheal medication accounts discuss these ideas at their television commercial brainstorming sessions. But that is precisely the point: These are revelations about the human condition that we don't want to hear. They are buried deep in our minds, deep enough that we are capable of forgetting them through the course of our daily lives. We do our best to ignore anything that reminds us of them.

All this philosophical theorizing leads to two basic points: Those of us with chronic illness of any sort are often reluctant to tell our stories, for fear of how they will impact the perceptions of those around us; those of us with inflammatory bowel diseases bear the extra burden of knowing that offering even the most simple, truthful explanation of our illnesses and their symptoms inspires discomfort and unease in our listeners.

So we learn, quickly enough, to keep our stories to ourselves. It is a lesson that sinks in, so much so that even when I am confident I will find in someone a sympathetic ear, I usually remain silent.

When we first moved to Worcester in August, our neighborhood had a block party, attended by at least a hundred of our new neighbors. This took place in late August or early September, during the time that I was just beginning to receive medicinal help for the disease, so I was still experiencing disease activity.

The party was one street over from ours, which meant that I had to walk two blocks back to our house when I needed a bathroom.

At one point during the party, while Anne and I were talking to several of our new neighbors, I suddenly felt some pressure on my bowels.

"I need to grab something at the house—I'll be right back," I said to the group, and excused myself.

Later, after I had returned, one of the neighbors with whom I had been speaking when this happened approached me.

"My sister has Crohn's disease," she said. "So how are you doing?"

I was taken aback. How did she know? I realized after a moment's thought that Anne must have said something about it to the group after I had excused myself.

Given this neighbor's obvious knowledge of the disease, and the likelihood that she would have some understanding of, and empathy for, my situation, I could have been honest with her easily. I wanted to be honest with her. But it was as if some barrier had formed in my mouth, a filter through which the truth could not escape. Instead, I pushed the "Play" button on the tape recorder:

"I'm doing fine," I said. "I have a pretty mild case of the disease."

Later, back at home, I chided Anne for what she had done.

"I would prefer to keep the disease to myself," I told her.

"Why? I didn't think it was a big deal."

"It is to me. I just would prefer to keep it to ourselves. I don't really like having to tell everyone all about it and explain everything. It's embarrassing. Just let me tell people if I decide I want to tell anyone."

"Okay," she said, with a shrug.

OF COURSE, while I would not allow myself—or my wife—to talk about my disease with anyone, I was desperate to do precisely that.

Like anyone with a chronic illness, at times I wanted to feel the sympathy of others. I wanted other people to understand what I was experiencing, to recognize the extent of my suffering, to offer me comfort and support. At times I wanted to share my sorrows with others so that I could learn if anyone else was undergoing similar experiences.

No matter how desperately I wanted these things, though, I could not budge that filter I had installed in my mind and my mouth, the one that either hid the disease from others or sugarcoated the realities of it.

I could not budge it myself; my stay in the hospital, fortunately, did it for me.

MY CONVERSATIONS with others who had Crohn's disease had been extremely limited up to that point. After I was diagnosed, my mother passed along the phone number of a family friend who had, upon her diagnosis many years ago, had a complete colectomy. I spoke with her once on the telephone, to get some advice about gaining weight after one of my flares. My wife's aunt and uncle lived next door to a woman with Crohn's; I had a very brief conversation with her once in the driveway of her home.

Beyond these experiences, all of my contact with fellow sufferers had been through print—books, magazines, the Internet.

While I was in the hospital, I was visited every eight hours by a technician who would take my blood pressure, check my temperature, and record the information on a chart. It was usually a different technician at each shift.

One afternoon, a day or two before I was released, a short, thin woman with long and dark curly hair knocked and wheeled in the familiar cart. This was her first visit to me.

"Are you the one with Crohn's disease?" she asked me. She had a thick Massachusetts accent.

I nodded.

"I have it, too."

"Really? Where do you have it?"

"Small intestine."

The symptoms of the disease can vary widely, depending upon which part of the digestive tract it strikes. Only around 20 percent of Crohn's patients have it exclusively in the colon, as I do; most people have it in the final third of their small intestines. For those people—unlike myself—abdominal pain is often the worst symptom of the disease.

This was the case with the technician, who took pain medication, and had had two surgeries to remove pieces of her small intestine that had been causing her significant problems.

I know this because she told me her story in the hospital room that day, and—for the first time—I told her mine. It must have been a combination of circumstances that loosened my lips: the fact that I knew she would understand my experiences; the fact that we were in a medical setting, even though she was not my doctor or nurse; and the fact that I had spent the last five days in a hospital bed, weakened physically and mentally, and anxious about my future.

By the time she left I felt, at least mentally, markedly better than I had before her arrival. I understood that this lift in my spirits resulted from our conversation, and from the opportunity I had to speak openly with another about my disease. I had been given a glimpse into her life with disease, and I had given her a glimpse into mine. Those glimpses helped me to recognize the value of our exchange, both the listening and the telling.

In the listening I heard a story of suffering comparable to my own. I heard about experiences against which I could measure my own, both favorably and unfavorably. Seeing my experiences in the light of another's was illuminating; it helped me see how normal and routine, for other Crohn's sufferers, was much of what seemed to me, in my isolation, so exceptional and unbearable.

In the telling I was able to shape my experiences, to put them in the form of a narrative that made sense to me. It is through stories that we make sense of our lives, and the experience of disease is no different. We are hard-wired, as human beings, to create order from the chaos of our daily experience through the narratives we tell each other about the courses of our lives. Those narratives place the events of our lives in meaningful sequences.

Telling my story in the hospital helped me begin to do that with my disease. Both in the telling itself, and in the reflecting upon that telling afterward, I began to see for the first time how I had been following a pattern of delaying treatment until it was too late, how my worst experiences with disease seemed to come after I had stalled and hesitated my way into serious flares. I began to wonder what other patterns might emerge from telling a more detailed version of my story, what other lessons I might learn myself and perhaps share with others. Lying in that hospital bed, it became clear to me for the first time that I had to construct this very narrative, the one you hold in your hands, to tell my story both for my own sake—to give you a glimpse into my darkness—but for the sake of others as well, to help them understand how that telling brings both comfort and meaning to an otherwise chaotic and meaningless experience.

And so in the weeks following my hospital stay, in addition to the slow work I began to do on this story, I became more open about my condition with those around me, beginning at school.

Early one morning I mustered my courage and sat down with my
department chair in her office to explain to her exactly what was
wrong with me. As other colleagues stopped me in the hallway to
check on my health, once I had returned to work, I let myself name
my condition to them, and offered simple explanations of the disease's most basic features.

It was this newfound openness, this willingness to tell my own
story, that led me to confront a version of my pre-hospital self, a
meeting that convinced me even more thoroughly how important
it was for me to tell my own story.

IN THE LATE SPRING, another professor in my department
approached me to discuss a student who had been having some
trouble in his class. This student had been missing classes recently,
and the quality of his work had suffered a marked decline from the
beginning of the semester to that point, which was just a week or
two away from finals.

Once he had called the student into his office to discuss his
work in the course, my colleague had discovered that the student
had been ill for much of the semester, and had been shuttling back
and forth between doctors who initially were baffled by his symptoms. Eventually, though, the possibility of Crohn's disease had
been broached to him, and by the time my colleague approached
me the tests that would help confirm that diagnosis had already
been performed.

In his conversation with the student, my colleague could tell
that he was in a state of complete emotional turbulence, and was
terrified at the possibility of this diagnosis. My colleague had
approached me to see whether I might meet with the student to
discuss his concerns, answer any questions he might have about the

disease, and provide a positive example of someone who was living with it.

Of course I agreed, and a few days later my colleague ushered a very large—both tall and stout—young man into my office. At first I thought there must be some mistake, because Crohn's disease almost always produces weight loss. I learned later that, in some rare cases, the disease—depending upon the precise location of the inflammation within the intestinal tract—can have the opposite effect, and can produce weight gain.

I invited him to sit down, and asked him to tell me his story.

"Well, I've been sick," he said. "And they think it's Crohn's disease."

"What sort of symptoms have you been having?"

"Stomach pain. Really terrible stomach pains, especially after I eat."

"Anything else?"

"Mostly the stomach pains."

"No diarrhea?"

"Oh, yeah—that too. After I eat."

"Anything else?"

"Mostly those things."

And so it went, for ten or fifteen minutes—me prompting him to tell me his story, and he refusing to offer me anything more than the most basic information about his condition, giving me even that somewhat reluctantly.

I found it maddeningly frustrating. Flush with my new awareness of the importance of sharing my story with others, I wanted him to tell me *his* story—to describe for me what his life had been like up to that point, to narrate the onset of his symptoms, and to describe his current condition in as much detail as possible. I wanted all of this information because it would have helped me to compare our experiences, and to decide what sorts of information, advice, and perspective I might most usefully give to him. But I also wanted to do him the service of offering him a sympathetic ear, a sounding board against which he could begin to organize the undoubtedly chaotic events of the past few months into some kind of order.

But his filters were already too firmly stuck in place. Eventually I gave up, and simply began to tell him everything about the disease that I wished I had been told after my own diagnosis. At the end of it, he said I had answered all of the questions he had, and he thanked me for my time. I told him to please get in touch with me again if he had any additional concerns, and to stop by my office at any time.

I haven't seen him since.

Of course I understood, even then, that he was behaving in a perfectly normal manner—behaving as I would have behaved in his position just a few months earlier. He understood that you do not discuss your malfunctioning body with strangers. He understood that you do not just open up and tell your disease story to anyone who asks how you are doing. And an eighteen-year-old college student, above all, does not sit in the office of a professor and talk—however persistently the professor questions him—about the frequency or consistency of his bowel movements.

My encounter with this student was perhaps the first instance in which I sought out the disease story of a stranger. Since then, I have undertaken this process more actively, and now seek out such stories wherever I can find them. I do so partly in order to provide me with new perspectives on living and coping with my own chronic illness, and partly to help others experience the same sense of satisfaction and release that has accompanied the telling of my own story.

No doubt I could have come to this lesson far more easily and quickly by taking advantage of a local Crohn's disease support group. One exists at my local hospital. I would certainly recommend to others, especially to those who may not be as inclined as I am to put their narratives on paper, to take advantage of these opportunities to tell their stories and hear the stories of others.

Support groups are simply not for me, and I suspect I am not alone in this respect. I am better at one-on-one conversations, or in classroom situations in which I can direct and control the conversation. I have a harder time finding my voice in the more unstructured and egalitarian setting of the support group.

So I will continue to put my own story down on paper, if for no other reason than to help me continue to make sense of it. I will seek out the stories of others in the best ways I know how to elicit them.

RECENTLY I WAS ATTENDING an academic conference with a friend whom I hadn't seen since I was diagnosed. I told him my story; when I was done, he told me his. He had a degenerative condition in his hands that prevented him from typing and writing, and he was slowly losing the strength of his grip. Though I had known him for several years, I had never known about this condition, and I commented upon that.

"We all carry around our secret pains," he said to me afterward, sighing.

Chronic illness has taught me to see, to understand, and to respect the secret pains of others. It has taught me, as well, to make my own pains a little less secret.

UNBURDENING MYSELF of my secret pains, though, remains difficult, however beneficial I know the telling of my story can be.

Part of that difficulty stems from the lingering reluctance I have to be labeled, and to air in public what at least some of my listeners no doubt would prefer that I keep private. But after five years with the disease, part of that difficulty now comes from the accumulation of emotion that has built up around the story.

Telling my story means, in a very small way, reliving experiences that have been extremely difficult for me, both physically and emotionally. Telling my story means rehashing events that I often wish I could forget.

The writing, editing, and even final proofreading of almost every chapter in this book has found me, at one point or another, sitting on the couch next to the computer in my basement, crying shamelessly—lamenting the past, fearing the future—and trying to forget everything I have been describing in this book. The pages of this manuscript have been stained, metaphorically and sometimes quite literally, with my tears.

What those tears are calling for is simple enough: They want me to put this year behind me, they want me to think only of the future, they want me to dream of a future in which this past will never have existed, will be like a nightmare from which I have now awakened. They want me to forget my story, and to begin my life anew.

But I cannot forget, however healthy I am. Cannot and must not.

I am learning to tell this story.

9

TIME PASSES

Learning to Be Healthy

[Summer 2001]

Learning to be healthy, like all of the learning experiences I have described in this book, has not been the sort of lesson I could achieve in a moment of revelation and then file away. Learning to be healthy has been an ongoing challenge for me, one that has stretched from the days following my hospital stay to the moment I am typing these words. Learning to be healthy has not gotten one bit easier, and I don't expect it to get any easier.

I can demonstrate this most clearly by describing how I learned to curb my daily drinking.

I HAVE STRUGGLED with alcohol since my very first taste of it— two or three bottles from a twelve-pack of Michelob Light, choked down in the backseat of a station wagon on the way to a U2 concert with Tony and some friends of ours during my sophomore year of high school. It was a night of firsts for the Lang brothers: I drank my first beer, and Tony smoked his first joint.

Tony's first contact with marijuana didn't make much of an impression on him and never developed into a habit. In this, as in many other ways in which Tony has preceded me in eventually shared experiences, I wish I had followed his wise example.

Shortly after that first taste of beer, I began drinking at every opportunity possible. Such opportunities were relatively infrequent for an academically motivated suburban kid who came from a large family with a mother who stayed at home and kept strict tabs on our comings and goings. But by my junior year of high school I was out most weekend nights, pouring as much beer down my gullet as I could between the time I left home and my curfew hour. I made friends with peers who had similar habits, and we encouraged each other in our shared vice.

In college, with parental supervision lifted, I drank a little bit more each semester, and carried my habits home with me on breaks. During the summer between my sophomore and junior years of college, this came to a head one Sunday morning after a particularly bad Saturday night.

I had come in so late that my parents had not bothered to wake me for church—an absolutely unprecedented event in our household. When they returned, my father summoned me out onto our back porch, and we sat on the steps, both of us looking out into the backyard. The sun was shining on a warm July day, and I could still feel the alcohol in my system, giving everything around me a sense of unreality. My father was not angry, as I expected him to be. He seemed confused, like a man confronted with a problem he simply did not know how to resolve.

"Jim," he said to me slowly. "Your mother and I would like you to give up drinking for a week."

"Okay," I said, grateful to be let off so easily.

And that was all they asked.

And that was all I did. I gave up drinking for one week. I told my friends that I wasn't going out that week, and I stayed in every night and read. That didn't satisfy my mother, who insisted that I needed to learn how to go out with my friends and still not drink. I knew that, however, was impossible. But I had still fulfilled their request to the letter. When the week was completed, I gradually resumed my old habits.

I suspect I could easily have lost control in a more serious way if I had not met Anne a month into my junior year of college. In the weeks before I met her, I remember going out with some friends for lunch on a Sunday after a long weekend of drinking, and ordering one of the restaurant's extra-large mugs of beer. The next thing I knew I was asleep on the couch in my room. I awoke to find a note, scrawled on a piece of cardboard and pinned to my chest, from the two girls I had been with, describing how they had to drag me home and pour me onto my dorm-room couch after I had begun nodding off at lunch. I remember going to the late-night Catholic mass in my dorm's chapel that evening, sitting by myself and wondering whether I had traveled so far down the road to alcoholism that I would need professional help getting back.

A few weeks later I met Anne, and my relationship with her became serious very quickly. Both she, and the importance of that relationship to me, helped temper the worst excesses of my drinking. I renewed my academic work with vigor, and studied my way back onto the dean's list that semester, after the mediocre grades of my sophomore year.

But if I kept better control of myself for the remainder of my college years, and in the early years of our marriage that followed, it didn't mean I drank any less. I simply drank more regularly. So regularly, in fact, that by the time I graduated from college and had moved on to graduate school, I had settled into the very regular and steady habit of drinking three beers every night before I went to bed.

That habit lasted from my senior year of college until my stay in the hospital, with perhaps—over the course of those ten years— two or three dozen days on which I did not have at least one alcoholic beverage to drink.

Of course I had no good reason to drink. My parents and siblings all loved me, I had material comforts, and I had lived a childhood of security and support that most people would envy. Neither of my parents drank excessively; my mother was, and remains, a near-teetotaller, while my father would have a few drinks at parties, or perhaps a drink when he came home from work. I was always an A student, with a strong love of literature and writing, and my academic success was praised and rewarded both at home and in school.

Thus any psychological angst I could claim to have driven me to drink would have been completely self-induced.

That doesn't make my struggles with alcohol, which continue to this day, any less real. Looking back, I am convinced that I settled on the habit of drinking three beers every night before bed as a means of keeping my drinking in check. As long as I had a strict routine, one that limited my daily intake to an amount that did not cause me either hangovers or serious health problems, I could continue to drink without losing control.

But the best evidence that this strategy does not preclude my being, at the very least, a problem drinker was that except for the absolute nadirs of my flare-ups, I continued to drink without interruption throughout all of my illnesses.

If I had no good reason to drink before the spring of 1996, I certainly felt justified in my drinking after that. I had a chronic illness, after all—I was suffering physically and mentally. Drinking helped me cope, drinking helped me forget, drinking helped me feel like I was twenty years old again, the owner of a healthy body that could withstand any sort of gastrointestinal pounding I pushed it to endure.

Never mind the fact that drinking three or more beers every day will loosen anyone's bowels a bit, especially mine, and never mind that the alcohol probably interfered with the effects of some of the medications I took. My gastroenterologists had assured me that nothing that passed between my lips could affect the course of the disease, and that assurance provided me with the excuse I needed to continue drinking.

If anything, the development of the disease entrenched me all the more firmly in my habit of capping off the day with those nightly beers. The arrival of my children, which had coincided with the onset of my disease, had pretty much reduced or eliminated those occasions on which I could stay up late into the evening drinking at parties or in bars, unconcerned about the effects such behavior would have on me when I woke the following afternoon. The severely limited opportunities I had for binge drinking intensified my commitment to my milder habit of nightly drinking.

I tend to latch onto routines in my private life and then stick to them tenaciously. My nightly drinking thus became wrapped up into the routine of relaxation and unwinding I had developed to help me finish the day.

During the semester I am often at work on reading or writing or preparing for class until eleven o'clock in the evening. At that point, I close my book or shut down the computer and crack my first beer. I flop onto the couch, turn the television to whatever sitcom happens to be in the syndication schedule at that hour, and

open the newspaper. I never have much energy for the news section; this is the point in the day when I am trying to relax my brain,
and I spend much more time poring over sports scores and reading
the comics than I do analyzing world politics.

The combination of the mindless banter of the television, the
mindless content of the paper, and the mind-numbing effects of
the alcohol gradually allow me to decompress, washing away the
concerns of the day. It takes an hour to drink all three of the beers
I have set aside for the evening, and by the time I am finished I am
exhausted and fully ready for sleep.

I DRANK THE LAST OF my nightly beers on Sunday, February
11, 2001. I was sick at that time, of course, in the mode of moderate disease activity that had been plaguing me since December. But
the next day, Monday, at around dinnertime, after I had eaten a
light dinner and had drunk a glass of Gatorade to keep myself
hydrated, I began experiencing the severe vomiting and diarrhea
that would eventually land me in the hospital.

I had no choice but to stop drinking at that point. I couldn't
keep any food in my stomach for several days, and by the time I
was able to keep anything down I was in the hospital. When I got
out of the hospital I was taking so many medications, and so much
prednisone, that I was afraid to drink.

But even at that point, I had every intention of returning to my
old habits. I left the hospital taking eighty milligrams of prednisone
a day, with instructions to reduce to forty milligrams (a dose I was
more accustomed to) within the next four weeks. I vowed to avoid
alcohol until I had reduced to that dose; I held out the promise to
myself that I would allow myself to drink again at that point.

Those four weeks were the longest I had ever gone without a

drink in my adult life. It was not as difficult for me as I would have expected, because I was so exhausted from the illness that I generally fell asleep as soon as I lay down on the couch after we had put the children to bed, and I crawled into bed before 10:00 PM on most nights. I did not need my evening routine to help me decompress and prepare for sleep; the disease was doing that for me.

But after those initial four weeks, I still had not reduced to forty milligrams of prednisone per day. That didn't happen for another two weeks. I planned my reduction in those final two weeks so that I would get to forty milligrams on a Friday; I would celebrate the weekend and my return to a familiar level of medication at the same time.

The day of reduction finally arrived, and my health was steadily improving. That evening I held off until my usual hour, 11:00 PM, and sat down on the couch with the first beer I had touched in six weeks. Six weeks. Watching the tiny bubbles foaming to the surface of the glass, I could hardly believe that it had been that long. I had not been dry for that long since my sophomore year of high school—fifteen years ago.

I reflected for a few minutes before I took that first sip. For the past few weeks, even as I recovered my strength and energy, I had been sleeping at night better than I used to. My three nightly drinks always knocked me out for the first few hours of the evening, but I usually woke up between 4:00 and 5:00 AM, dehydrated, and then drifted in and out of sleep until it was time to get up. Without the alcohol, I found myself sleeping until something besides my own body—an alarm clock, the girls, or a cat—woke me.

For most of my adult life, I had yearned for a nap every day. Most people want to nod off when the body's metabolism slows down in the early afternoon; I was ready to head back to bed at any point during the day. Since I had not been drinking, I found that I had more energy during the day. I no longer begged Anne to let me catch a quick nap on Saturday and Sunday afternoons, and I was no longer tempted to sleep for thirty minutes between classes in the easy chair in my office. I had energy during the day, and was tired enough in the evening that I fell asleep without much trouble.

I found I actually began to enjoy the mornings when I had not drunk alcohol the evening before. I did not snap at the children, I had an appetite for breakfast, and I had better control of my bowels. The morning had been transformed from a time period through which I tended to sleepwalk to one of the parts of the day that I actually enjoyed.

These changes had clearly improved the quality of my life; I could see that easily. Did I really want to take that sip?

Of course I wanted it, and I took it. It tasted funny to me, but I choked it down, and drank the two more I had brought out after it. By the time I was finishing the third one, it tasted a whole lot better.

I couldn't hold off that Friday evening; I had built that moment up—my triumphant return to alcohol—for too long in my mind. But that evening I fell into my customary sleep pattern: I awoke at 4:30 AM, thirsty, and got a glass of water. I tossed and turned in bed for two hours, and drifted off to sleep again sometime between 6:30 and 7:00 AM—just about an hour before I had to get up and get Katie off to her swimming lessons. That day, for the first time in a while, I sneaked upstairs for a nap while Anne took the girls out shopping. I was tired.

I found myself unable to resist the lure again on Saturday night; the association of alcohol and weekend nights was simply too strong for me to alter. But I did manage to scale back to just one beer on Sunday evening. On Monday, Tuesday, and Wednesday I had none.

For the next few weeks, in April and early May, I maintained this pattern. I kept myself dry Sunday through Wednesday; I allowed myself a beer or two on Thursdays, and I reverted to my old habits on Friday and Saturday.

In the meantime, I was slowly dropping the prednisone—from forty milligrams a day to thirty, to twenty, and then in slower increments to fifteen, ten, five. I had gradually eliminated Cipro from the medications I was taking. My bowel health was taking on a new character, and it was evident now that the Imuran had begun working. I would have three or four bowel movements in the morning, all within an hour of waking, and all diarrhea—but then I would be done for the day. I could actually spend the majority of my day living my life and not have to worry about being continually on the watch for restrooms. I no longer saw any blood in the toilet, and my strength and energy were now back to where they had been the previous summer.

In May I saw my gastroenterologist and described what I had been experiencing. He reduced the amount of Asacol I was taking every day from twelve pills to eight, and this cleared up my diarrhea within the space of a week.

The best part was that school was finishing, and I had the summer before me—the first summer I would have, since high school, in which I was not working at least part-time, and to which I could devote myself entirely to my two great joys: my family and my writing. I would also be able to relax my mind and release the stress that had been accumulating in my brain for the past nine months of my freshman year as an assistant professor, and for the past nine months of illness.

I could hardly allow myself to open my eyes to it, after the nine months I had lived through, but as school wrapped up in the third week of May, and spring was finishing its preparations for what promised to be a glorious summer, I was beginning to believe that a new era of remission and good health was on the horizon.

And that, of course, is precisely when the headaches and fevers began.

I WAS CERTAIN, at first, that I had a cavity. I initially became aware of a problem when I began to get sharp headaches, with the pain localized in my upper jaw or behind my eye. They started off slowly, but then worsened over the course of two weeks to the point that I had to take aspirin at night to get to sleep.

At the same time, I developed a continuous, low-grade fever. These fevers never reached the levels of the fevers I had experienced in the period before my diagnosis; at most, they would peak at 102 degrees. Usually, they stayed in the range of 100–101. Like all fevers, they were worse in the afternoon and early evening. I became obsessed with monitoring my temperature, and took it constantly. Typically I would be fine until at least lunchtime; by 2:00 PM the fever would be gradually moving into the 100-degree range, and it reached its peak between 4:00 and 6:00 PM.

Was it possible that a cavity could cause a fever? I reasoned myself into believing it was. I generally carried what I learned from my disease about doctors and medicines into other areas of medical care, and one day I had spent some time asking my dentist about teeth and their care. He had explained the basics of his profession to me eagerly, as if he had been waiting for someone to ask him just these sorts of questions, and one of the points he had made was that tooth decay was actually a bacterial infection.

Pain in my upper jaw, a fever to fight off an infection—massive tooth decay, right? The problem was that I had no other symptoms of a cavity: none of the teeth in the upper jaw were sensitive to the touch, and hot and cold liquids did not increase the pain in any way. The dentist I saw two weeks into this new problem pointed all this out to me, though I had known it myself, and suggested that my problem could involve the sinuses. He suggested that I consult my regular physician to try an antibiotic for a sinus infection.

That I did, and Dr. Honig seconded the diagnosis and prescribed an antibiotic. It didn't work. After ten days on that antibiotic, we tried a different one. It didn't work either. At this point nearly five weeks had passed, and I was still having the same regular fevers and head pain.

Fortunately, summer had arrived, and I was finished with school. Anne and Katie still had another six weeks of elementary school, and Madeleine was going to her sitter for six hours during the day so I could write. With the help of aspirin, I could write through most mornings. But I had little energy, and I was starting to get worried.

In the back of my mind loomed the diagnosis that I was becoming more and more concerned about, given that I was having unexplained head pain: a brain tumor of some sort. I did not discuss this fear with anyone, but as the days passed and the antibiotics failed to help, it seemed more and more likely to me.

Finally, late one evening I did an Internet search on brain tumors, and read as much as I could find. I was distressed to learn that headaches and fevers can, indeed, be common symptoms of brain tumors; I was somewhat relieved to discover that usually other symptoms accompany brain tumors—neurological symptoms like loss of coordination or vision problems. Still, I knew better than anyone how the body does not conform to what medical books—or Internet sites—say that it should be doing, and so the possibility remained in my mind.

Dr. Honig's eventual recommendation that I get a CAT scan of my sinuses did not help matters any. I wondered if he were using the language of investigating my sinuses in order to cover the fact that he was checking for head tumors of some sort.

My CAT scan appointment was at 7:00 one late June morning. I arrived on time and was immediately ushered into a room that contained a massive white ring, through which a bed on a track could pass. The nurse strapped me onto this bed, and then I was slowly jerked and pulled into position inside the ring. Somewhere inside that ring, something was spinning; I could hear the machine whirling around me, taking pictures of the inside of my skull in a way that I could not understand. The nurse told me not to move; I did my best to lie completely still. Of course I was worried that I might have to run to the bathroom at any moment, but fortunately this new problem—whatever it was—had not activated the disease in any way.

Suspended in the ring, eyes closed, the smell of medicine in my nostrils and the clicking and whirring of the scanner in my ears—all of it immediately brought me back to the hospital, and I thought of how easily I could be returned to that environment by an uncooperative body. I felt small and frail and out of control, a weak and sick body being processed through a machine, once again caught in the grip of the medical establishment in which I had little faith remaining.

It was over in fifteen minutes, and I was sent on my way, told that someone in the lab would read the results and convey them to my doctor. If they noticed something amiss, they would let him know immediately; if the results were routine, it might take longer for them to file the report.

So I found myself for a second time in a holding pattern, waiting nearly a week for test results that would define a condition that was interfering with my life in a significant way. I was not able to wait as long as I apparently was supposed to; I called the office a day or two early to see if the results had come in.

When Dr. Honig called me back later that same day he had news which, whatever he may have thought of it, seemed good to me: I had sinusitis in one sinus, and what looked like either inflammation or a polyp in the other. The word polyp initially shot up my cancer radar, but I was assured that sinus polyps were a common and benign condition. Later that week, while golfing with a neighbor who was a physician, I grilled him on my diagnosis and received confirmation of this.

If the sinuses continued not to respond to antibiotic treatment, eventually we would have to consider surgery. But Dr. Honig wanted to try one more kind of antibiotic before he sent me to a specialist.

And lo and behold, that last-chance antibiotic did the trick. Within a week of starting this final antibiotic the fevers had disappeared, and the headaches and facial pain were slowly receding. By the time I finished the antibiotic, and all of my symptoms had completely disappeared, six weeks of summer had passed me by.

Throughout those six weeks I had been forced to cut back on my drinking even further. I had noticed one day that my aspirin bottle contained a warning that people who regularly drink three or more alcoholic beverages a day should consult their physician before using any sort of pain reliever. It was a warning I would have ignored a year earlier; now, I decided it was best for me to just stop drinking so I could continue to take the aspirin that relieved the headaches and fever.

Once I had finished the antibiotic and felt myself cured of the sinus infection—although I was told that it could recur—I returned to my more limited drinking habits. I allowed myself beer on the weekends, but did my best to stay dry on weekdays. This was especially difficult during the summer, when I had no classes or office hours to get up for in the morning, but I forced myself to stick with my new regimen.

Put quite simply, what helped me in my resolve to cut back on my drinking was my rediscovery of the pleasures of living.

Roll Tape: Scenes in a Life Recovered

SCENE: Concord, Massachusetts, June. We have driven to this historic city outside of Boston and rented a canoe for a trip down the Concord River.

Anne and I situate ourselves at the ends of the canoe, with Katie and Madeleine, dwarfed by their oversized orange life preservers, on the benches between us. We push off from the dock and slowly paddle our way out into the river, toward our destination a mile or two down the river.

Though we started out with several other canoes, we round a bend and find ourselves alone, in the quiet of a short and isolated stretch of the river. Vegetation lines the banks on both sides of us, through which we can see houses in the distance. Suddenly, as if emerging from the river itself, a goose and six little goslings are paddling their way down the river beside us, perhaps a dozen feet from our canoe. We all notice them at once, and watch them in silence. The goslings paddle along quickly, rocking back and forth, behind their more smoothly gliding mother. Our canoe travels faster than they do; we slide past them quietly, and resume our paddling.

SCENE: Mid-morning, early July, on Chicago's Navy Pier, a promontory jutting into Lake Michigan from the heart of the city.

Anne, Katie, Madeleine, and I climb aboard the small red cars that will circle to the top of the 150-foot Ferris wheel—my sister, Peggy, who has accompanied us to the pier this morning, has decided to sit this one out. The wheel turns slowly, and we rise by inches to a place far above the pier, suspended and swaying lightly in the wind. I have a slight fear of heights, and this little gondola couldn't be more exposed. My heart beats quickly, and two or three small waves of panic come and go. I am gripping Katie fiercely by my side.

At the peak of the arc much of downtown Chicago is visible on my left; to the right, Lake Michigan stretches off into the horizon. The sky is clear, the day is warm, and the pier has begun to fill up with tourists like ourselves. They stroll casually around below us, some of them pointing up to the very car in which we sit. Out in the lake, I spot a yellow boat slipping over the waves, and I am hit with a desire to be on the water.

Down on the pier again, I buy us tickets for the *Sea Dog*—they are expensive, aimed at the captive market of tourists on the pier for the day, but I don't care—and we board the seventy-foot speedboat for a thirty-minute tour of the lakefront. Despite its large size, the boat reaches speeds of twenty-five knots, shooting spray into our faces and creating gusts of winds that feel like they will blow us right off the boat. This time it is Madeleine I am gripping tightly to my side. We rip across the surface of the lake, bouncing and sinking with the small waves, and I feel a powerful sense of freedom.

SCENE: Mid-July, a farm and vacation home in suburban St. Louis, owned by the family of Anne's sister's new boyfriend, Bruce, who both comes from money and has made a bunch of his own. We are in St. Louis on our three-week midwestern road trip, and have been invited out to the farm for a day of swimming, horseback riding, and relaxing.

After lunch and an early afternoon of swimming, Bruce rides from the farm up to the house on an ATV four-wheeler, one of those small, jeeplike vehicles that one steers and brakes like a bicycle, at the handlebars. He gives me directions to a pond we can see down in the valley below—their property stretches for acres in every direction—and I put Madeleine on the seat in front of me and take off, driving carefully over the path and out into the meadow that borders the pond.

Madeleine and I dismount and approach the edge of the pond, where we see tiny frogs hopping all around us. I catch one up in my hand and we admire it for just a moment before it hops away. We wander around at the edge of the pond, Madeleine chasing frogs and me watching Madeleine.

We return to the ATV and begin driving back across the meadow, but I stop at the sight of a greenish bump just off to one side of us. Sure enough, we have discovered a medium-sized box turtle. I pick him up and we watch him react, curling into his shell. Together we gently stroke the part of his back leg that remains exposed, feeling his tough, leathery skin. We admire the markings on his back.

Madeleine wants to bring him back to the house with us, but I say no.

"Let's leave him here," I say. "He would miss his friends and his home."

SCENE: Nantucket Island, thirty miles off Cape Cod in the Atlantic Ocean. Early August. Warm summer morning.

I load my bike and Katie's bike into our rented car, and Anne drives us from the vacation home we are renting on the island with another couple to a bike trail a half mile away. The road to the bike trail is narrow and usually filled with cars, and I don't trust Katie's fledgling bicycle-riding skills enough to let her negotiate the road alone.

The bike trail is wide and smooth, and crowded with bikers and walkers heading to and from the beach that sits about a mile from where we are donning our helmets and climbing onto our bikes. Katie starts off in the lead, pedaling hard to reach the speed she has decided is right for a bike ride with your dad. She weaves slightly back and forth when the trail is open, but when we approach other people or bikes she seems able to maintain her control, moving carefully to the edge to avoid collisions.

At the beach we put out bikes in the racks and I buy snow cones at the concession stand. We sit and watch people walking down the long sandy strip to the ocean.

Back on the bike trail again, I watch her slowly becoming more confident on her bike. She has just learned to ride by herself the week before this vacation, and she rides for the pure pleasure of it—for the joy of practicing and mastering a hard-won skill. I pedal slowly behind her, straining my ears to catch the constant stream of chatter that she keeps up throughout the ride. I don't hear most of it; I am content simply to watch and hear the sound of her voice.

"Okay," I keep shouting ahead. "Keep your eyes on the trail."

SCENE: An Audubon Society wildlife sanctuary, in Princeton, Massachusetts, just about twenty-five minutes from our home in Worcester. Mid-August. A thousand acres of hiking trails through meadows, woods, and ponds.

I searched this place out on the Internet, and brought Katie and Madeleine here one Friday afternoon. When we pull into the parking lot, we see sheep roaming around us. We stick a few dollars in the donation box, grab a trail map, and head off toward Beaver Pond. I am carrying Katie's school backpack—pink, with a picture of Snow White on the back—filled with snacks and sandwiches and the girls' wading boots.

We trek happily across a meadow and into the woods, up and down the sanctuary's rolling terrain until we arrive at the pond. The trail takes us to within a dozen feet of the beavers' huge lodge, packed in with mud and sticks and small trees. The girls clamor for their boots, and wade out into the shallow waters. They are ostensibly searching for frogs, but in reality they simply like the fact that they can put their feet into the water without getting into trouble.

We sit on the benches by the edge of the trail and eat our lunches, and then set off again. Their little legs are holding out well; they are exhilarated to find so much to see, to climb over and under, to run up and down. We walk a mile and a half to a tremendous boulder that, as the nearby plaque tells us, was pushed to this spot by a glacier some 15,000 years ago. We climb partway up it, and scream out "Hallos" to the woods around us, listening for the echoes. The girls want me to climb to the top of the boulder, which would require at least partially climbing a small tree. I am tempted, but then I remember I am thirty-two years old, and that I am here with my young children, and I decide against it.

I'll be back another day.

IN THE THREE MONTHS of that summer, I lived more intensely than I had in any other period of my life. I seized every possible new experience that crossed my path, devouring each one of them greedily.

I learned to be healthy that summer by giving up the daily

drinking that had held me for ten years and by beginning to exercise regularly and watch my diet more closely. I hope that I can retain all these habits.

But I think I can sum up what I learned from that summer in two simple lessons. When illness strikes, in any form, rest and make whatever sacrifices are necessary to return to health.

When health returns, live.

10

LEARNING HOPE, CAUTION, AND BALANCE

A Return to Daily Life

[August 2001]

Our destination is Good Harbor Beach in Gloucester, Massachusetts, approximately seventy-five miles, as the crow flies, from our home in Worcester. The journey, I have been assured by a neighbor who takes his children there regularly, will take us at least an hour and forty-five minutes. Packed into our minivan are Anne and me, Katie and Madeleine, and Mike, a fellow professor in the English department. Mike and I were hired at the same time by the college; we are office neighbors, fellow writers, and friends.

The end of August is approaching. The days have been long and beautiful, and promise to remain so for another few weeks at least. School holidays are about complete, and I have been slowly,

163

sluggishly, turning my thoughts to school and to the upcoming year. I have been in the office a time or two already, preparing syllabi and doing some cleaning of my files. But I have been reluctant to really face the prospect of return. I am, first, mildly anxious that the return to school and work, with the stress that such a return will invariably create, will jeopardize my health again. But even more than that, I have enjoyed this summer with my wife and children as much as I can remember enjoying any time in my life, and I simply don't want it to end.

I banish thoughts of school from my head as we load up the car with blankets, beach toys, a cooler full of snacks and drinks, and extra clothes in preparation for our scheduled 9:00 AM departure.

I still don't enjoy mornings all that much, though they are no longer the trial for me that they had been just six months ago. I have settled into a happy routine: one or two bowel movements in the morning, shortly after awakening, and then no more for the rest of the day. Even as recently as March and April I would not have believed such a routine possible for me.

Minivan loaded, we pack into our seats and head out. We pick up Mike and are on the highway by 9:15, an amazingly timely start for us.

Mike, a bachelor, once made the mistake of cupping his hand and nipping at the girls' legs, pretending that he was a biting fish. He did not realize how quickly children grasp, and how long they hold onto, such familiar and easy games. Soon after we set off the girls begin giggling and protesting as loudly and unconvincingly as they can that they don't want any fish biting today.

Anne and I exchange looks and laugh, half at the children's simplistic duplicity, and half out of pity for Mike, who has sealed his fate with our children as the "Fishy" guy. He obliges them for several minutes, until Anne mercifully interrupts and offers the girls several choices of music on the car ride.

We sail along to the melodious strains of the soundtrack to the movie *Shrek*, and have made it almost an hour into our journey when Madeleine informs us that she has to go to the bathroom. A part of me wants to do a little celebration—for once, we are making an emergency bathroom stop for someone other than me.

I pull off at the next exit, and while Anne and Mike—rookies—scan the terrain for gas stations, I spot the likeliest source for an available public restroom in a nondescript brick building that has a sign outside labeling it a medical office center.

Number one source for public restrooms in an emergency: medical facilities of any sort. They are usually spotlessly clean, and—although I have never been questioned—I presume they would show understanding for anyone needing a bathroom as a result of an IBD emergency. Second best source: large hotels. They have no way to keep track of who belongs in the hotel, and public restrooms are always available somewhere in the lobby. If questioned, you are meeting a friend staying at the hotel. Third best source: fast-food restaurants. The teenage kids behind the counter at McDonald's don't care whether or not you have actually purchased a burger and fries before you use their bathroom.

So I pull into the parking lot, assuring Anne she will find what Madeleine needs inside. Anne doesn't believe me, but she leads Madeleine away by the hand and returns in a few minutes with a small grudging smile on her face.

"I guess I should trust the expert on bathrooms," she says, as she straps Madeleine back into her car seat.

In another forty-five minutes we are driving slowly through traffic in the town of Gloucester, searching for Good Harbor beach. After a stop at a visitor information center, we find it and, arms laden with beach supplies, weave our way through the crush of bodies in the sand and settle into a spot ten yards from the water's edge.

The girls want to go swimming, of course. They love the water, as I do, and will hop into a pool, a lake, or the ocean at every possible chance. So Anne and Mike settle down to spread out our stuff while I follow the girls, who run flapping through the sand, down to the water. I wade into the shallows just behind them.

Oh my God.

I have never felt colder water in my life. I feel like I have stepped into an icy cooler full of beer and soda at someone's back-yard barbecue. Within seconds of placing my feet into the water, my toes start to hurt. The girls have walked just a few feet ahead of me, and for a moment I wonder whether their love for the water will overcome their basic human need to keep their bodies warm. I have seen them jump and swim happily in hotel pools that to me seemed frigid.

But I am not imagining this. They turn to me, smiling but sur-prised, lifting their feet out of the water and squealing. They dash back onto the sand, and then make little forays into the breaking waves, perhaps hoping that the water washing into the shore will bring warmer temperatures.

"Girls," I tell them, "I don't think we're going to be doing much swimming today."

We are all disappointed, perhaps me most thoroughly. They will find other ways to entertain themselves at the beach, and they are probably convinced that the water will warm up eventually. I know better, and I know too how difficult it will be to occupy their attention for an entire day with sand castles.

But we have driven all the way here, and we have to make the best of it, so we settle down a dozen feet from the shoreline and begin building: a moat, surrounded by a great wall; a pyramid in the center covered with rocks, towers at its four base corners. We work diligently for a little while, but their attention begins to wander, and they occasionally make little forays into the water, still expecting to find the water temperature more friendly.

After an hour or so of sand play I decide that I have to brave the water myself, and so I take off my shirt, inform Anne that she is in charge of the girls, and step into the water.

Oh my God.

This time I do my best to shake off the numbing pain in my toes, and move cautiously out into the deeper waters. The ice crawls slowly up my leg: shin, knee, up my thighs and then pauses just below my waist. I am not sure I can stand the agonizing creep of the ice up my genitals, so I have reached the point at which I need either to retreat or to plunge into the surf, submerging the upper half of my body in one fell swoop.

Last summer I would have retreated. I would have seen no good reason to continue to endure the pain in my feet, for the clamp of the icy water around my thighs, for the whole silly experience.

This summer I am a different person. I am hungry now, hungry for experience and sensation in a way that I have never been in my life. I am healthy; I am free of the disease's constraints for the moment, and I want to cram all the life I can into this period of health, because I know it cannot last.

Earlier this past summer, at a swimming pool with some friends in Chicago, I had climbed up thirty feet of winding stairs to a ten-meter diving platform, dizzy and scared, and forced myself to walk off the edge of it and plummet into the diving pool below. As I stood a few feet back from the edge of the platform, I thought my heart would burst out of my chest, or that I would succumb to the weakness in my legs and fall to the platform floor. I could not even look down; I walked off the edge as if I were stepping off a curb and did not see the water until I opened my eyes ten feet deep into it, feet and hands painful and stinging from slapping the water so forcefully.

At the surface I could not help but laugh; it had been terrifying and painful, but I was glad to have done it. I could not remember the last time my nerves and senses had been so intensely alert—the last time I had felt so purely alive.

Standing in the surf, I recollect that moment at the top of the platform, and make my decision. I bend slightly at the knees, and leap into an oncoming wave.

I am confused at first, because I believe someone—someone who must have been waiting for me under the water—has taken a board and slapped me on the chest and face with it. The wind has been knocked out of me, I am gasping for breath, and I can feel the numbing cold, the aftereffects of the slap of that wooden board, on every inch of the surface of my body.

I stand quickly, the water just above my waist, and regain my senses. No one has slapped me; I dove into a sheet of icy water, and it took my breath away. I lower myself into the water to my shoulders, expecting to feel some diminution of the temperature shock the second time around. If there is diminution, I don't feel it.

However hungry I am for new experiences, I have had enough of this one. I turn and walk to the beach, my pace quickening as I near the water's edge. The girls are still at it in the sand, filling buckets with the frigid water and vainly attempting to create a moat, seemingly unfazed by the rapidity with which the sand soaks up each bucketful.

"Grab your buckets," I say, "and let's see what we can find along the beach."

I lead them down the beach, toward the high outcropping of rocks at the far end, a picturesque house perched at their pinnacle.

The shells are few and far between; we run into other families with the same intention we have, and there are simply too many collectors and not enough crustaceans. But we get lucky occasionally. The girls are not as picky as I am; I want them only to collect the best shells, perfectly formed. They are content with scraps and pieces, half-shells and discarded crab legs.

As we approach the end of the beach area, I notice that a small tidal creek runs along the edge of the beach into the sea from a salt marsh back inland. I shepherd the girls toward it. Along the creek I see two boys fishing in the water with nets, and I am curious. I approach them, and kneel down to see what they are seeking and what they have found.

"What're you guys doing?" I ask them.

"Catching crabs," they say. "See?" One of them points me into the bucket they have, which is indeed filled with tiny green crabs, climbing around in a clump of sea vegetation.

I watch them catch the crabs. They sweep their nets into the long strands of green and reddish vegetation that, I now see, fills the edges of the creek. They dump the contents of their nets onto the sand behind them, and invariably they find these little crabs crawling around in the stuff. They grab the crabs in their fat little fingers and deposit them in their buckets.

"Girls! Girls!" I shout to Katie and Madeleine, who are thirty feet away, digging in the shallow surf for something. "Come here, quick!"

They come sprinting over, impressed by the excitement in my voice.

"Look," I say, pointing into the bucket. Just then, one of the boys drops a newly caught load of vegetation onto the sand, and we watch as he sifts through it, searching for crabs. We see an abundance of other creatures there as well: tiny, swimming, millipede-like crustaceans, who crawl around frantically in the leafy material.

Madeleine squats over the vegetation and fingers one of the little creatures intently; I want to warn her that they might bite, but I settle for a more generic warning: "Be careful." I don't want to scare her away from animals, and, besides, they don't look like biters to me.

Katie, in the meantime, can hardly contain her excitement; she is literally hopping up and down as she tugs on my arm.

"Can we get a net, Dad? Did we bring any nets?"

We don't have any nets, so instead we attach ourselves to the boys, becoming their helpers as they continue to forage for marine life in the creek. Katie alternates between tending to the bucket, watching the little crabs, and encouraging the boys in their netting forays. Madeleine is not sure which interests her more, either: the process of catching things in a net, or the animals they find.

I spy Anne walking toward us from the other end of the beach, and I run to within shouting distance.

"Get the stuff!" I yell to her. "Bring it all down here!"

She can see the buckets and nets, and can guess the source of the excitement, so she and Mike dutifully collect our belongings and bring them down by the corner of the beach between the creek and the water line.

In the meantime—and it happens so gradually that it takes me close to an hour to realize it—the tide is going out. The creek has lost five or six feet of width, and at least a foot or two of depth. I step into the edge of it and notice, as well, that it has become warmer, at least at the edges. A cold current still flows through the deepest part of it at the center, but the edges are warm and pleasant.

The girls and I now begin foraging on our own. We grab clumps of the floating vegetation and pull it from the water onto the beach, where we make our own discoveries of tiny crabs and other creatures. Anne has brought us our own buckets, and we fill them with our finds.

One of the clutches of vegetation we pull up brings an incredible treasure: a live starfish. We can't believe our good fortune. I pick it up gently, and find it soft and knobby, yielding at least the surface of its body to the touch. It reacts slowly to our hands, curling its legs around the underside of my hand as I hold it in my palm. I pass it gently to Katie and Madeleine, who stroke it and admire it and then deposit it into our bucket, more excited about the prospect of other new animals than they are about this actual new animal.

The tide has now moved far out to sea. The creek is no more than two or three feet deep in the center, and perhaps a half-dozen feet across in most places. I notice that the water level has sunk below the stony ground at the base of the high rock outcroppings that mark the end of the beach, and tidal pools are forming. I notice, too, that this area of the beach has come alive with people, most of them small children with buckets, supervised by parents who help them find and collect the small animals that are scuttling about in their suddenly exposed environment.

I take the girls' hands and pull them through the creek over to the rocky tidal pools; Mike and Anne join us over there, and we see now that we have happened upon a treasure trove of collectible creatures. With the exception of our starfish friend, we had been pretty limited to small, flat-backed, greenish crabs, and those tiny millipede-like creatures. In these tidal pools we have more than doubled the number of species in our range: we find small, almost transparent, shrimplike creatures, hiding in the sand and darting away swiftly at any disturbance in the water; hermit crabs, from fingernail-sized to as large as an inch or two across the shell; snails and mollusks, clinging fiercely to the wet rocks; and several varieties of small fish, flashing back and forth in the small spaces of the tidal pools, seemingly in search of exits to the sea that the receding tide has closed to them.

Initially I help Katie and Madeleine with their collecting, but eventually I wander away from them and let Anne work with the girls. Mike has strolled off down the creek toward the ocean, and I am left to search through the rocks by myself. Ostensibly I am in search of another starfish; we had shown our earlier one to the two boys, and somehow it found its way into their bucket as they were leaving.

In reality, I am consumed by the same fascination with this eco-system that has infected my daughters. Growing up in the Midwest, I never had such easy exposure to wildlife, though I was fascinated by animals well into my teenage years. My early career dreams, before I understood how academic disciplines like chemistry and physics stood in my way, centered upon fields like veterinary medicine, marine biology, or herpetology (I had a long-standing and inexplicable interest in snakes).

That fascination with the natural world faded and nearly disappeared while I was in college, and throughout most of my twenties. So intent was I upon the world of literature and writing and thinking that I often ignored the external world around me, especially the natural world. Seven years in urban and suburban Chicago, too, throughout graduate school and the first three years of my academic career, limited the possible exposure I might have had to wildlife.

But in recent months I have felt that childhood fascination with the natural world resurge forcefully. The ground for that resurgence had been slowly prepared over the past several years, as my daughters reached the age at which most small children, I think, become fascinated with the other living creatures that surround us in this world. I couldn't help but catch part of their enthusiasm for the natural world.

After my stay in the hospital, in the months that followed and eventually opened up into summer, my interest in nature increased exponentially. I began to take the children on nature walks to the various parks around where we lived; I bought the first field guides I had ever owned, and carted them with me on bike trips through the woods, or on hikes in the local Audubon Society wildlife preserves, or on trips just like this one, to the local beaches. I bought and read nonfiction books on the natural world, and began watching nature specials on television in the evenings with Anne and the girls.

But I have not yet had time to reflect upon what has spurred this interest. Crouched among the tidal pools, some understanding of that begins to press its way to the surface of my mind. I reach into a small tidal pool and pick up a green crab, careful to avoid the small pincers that probably couldn't do me much harm, and set it on a rock a few feet away from the water. The crab momentarily searches for a rock to back under, but then changes tactics and turns hurriedly toward another pool, just a foot or so behind it. The crab's movements seem awkward, jerky—he is clearly more at home in the water—but they impel him along nonetheless, and he finds his watery destination.

Twenty feet away from me Anne leans over to touch and observe something Katie holds in her hand, while Madeleine squats beside them, trolling her finger through a tidal pool. Mike, a nature lover of much longer vintage than myself, has wandered up the creek into the ocean water, following his own interests in this world at the edge of the sea. Around me children are everywhere scrambling over rocks, splashing their feet in small pools, and shouting excitedly to the parents they are towing behind them. At my feet crabs scuttle about in search of shelter and food, shrimp and fish dart and flash in the sun, snails and mollusks suck diligently at the moss-covered rocks in the water.

Everywhere, life.

One could stand at the edge of this ocean, in complete solitude, and see only the gentle crashing of the surf against the sand. But come on a sunny summer day, take just a moment to peer into these waters, and life appears everywhere—life animal, life plant, life human; life multicolored, multiformed, multimotile; life that swims, crawls, hops, scuttles, clings, walks, oozes.

Life that has existed on this planet for hundreds of millions of years, that has existed in some of these very same forms for at least that long, and that will continue to exist in these same forms—we hope—for many more hundreds of millions of years. Life that multiplies, life with appetites, life that senses, life that seeks to survive at all costs, against all odds. Life that possesses such infinite variety and diversity it can exceed our capacity either to understand it or to appreciate its beauty.

I am no wide-eyed, naive observer of nature. Stay here long enough, and you will see these creatures devour one another alive; follow enough of these children around and you will find one of them pulling the claws off of every crab he can find; cut open these snails and you will find parasites that have drilled into their shells and curled deep within their intestines, siphoning off their food.

But even these aspects of life are part of the cycle, and do not slow or deter nature's creatures from their basic drives to produce and reproduce new life. My still superficial readings in the literature of the natural world have conveyed one lesson to me more clearly than anything else: Living things seek, first and foremost, survival and reproduction. The goal of life is to produce more life—however scarred, however transformed, however sickened by that life's encounters with this world, with its own genetic defects, and with its own mortality. Life pulses on.

In my understanding of that fundamental fact, in the basic evidence of it that I have come to see in the natural world around me, in my reflections on it—on a warm August day, crouched in the tidal pools of Gloucester, Massachusetts, with my wife and children and my friend—I have begun to take comfort. I have begun to see my own life's place within these larger cycles, and to see connections between my struggle for life and health and the struggles for life and health of so many of earth's creatures around me.

Between the sea and the sky, my feet moistened in the water and my back baked by the sun, I feel the force of life pulsing within me, too. I feel strong and resilient, the survivor of a year in which my body has ravaged itself for no reason that anyone can fully understand. I have shat and bled and wept my way through a terrible year, but I have scratched and clawed my way to this moment of health and happiness, and I have my wife and my children and my friends and my God intact along with me.

Whatever happens to me in two weeks, in two months, in two years, I have this moment to hang onto, this moment of both joy and wisdom. Without having experienced what I have undergone this year, I am certain I would let this moment slip away— unnoticed, unmarked, and unremembered. Instead, it is etching its way into my body and my brain, and it is a moment to which I know I shall return, again and again, in the years to come.

THE SUN HAS BEGUN its descent toward the horizon, and we want to spend an hour or two wandering through the tents and makeshift shops of the Gloucester Waterfront Festival, a mile or so from the beach. Reluctantly, the girls spill their buckets into the tidal pools, and we watch as the small animals find their way back to their homes in the water and the waves.

Slowly we return to our blankets and cooler, and slowly we gather everything up and trek over the sand back to our car. We cleanse ourselves briefly in the beach's outdoor showers, and pack our belongings into the minivan.

At the Waterfront Festival we wander in and out of the tents that are selling an incredible variety of handmade gifts and household items, occasionally purchasing the things that catch our eyes: Mike buys some pieces of driftwood painted with New England lighthouses, and Anne and I pick out some dried starfishes for the girls. They will end up atop the bookshelves in our family room, where they sit to this day. Further along, we let the girls fill small bags with the colorful shells an older woman is selling; useful material, I am certain, for a future project.

Tired of walking, we return to the car and drive along the harbor until we spot a restaurant that looks promising: Mr. T's Lobster House. Inside we settle down and I order a beer. It tastes good, absolutely perfect after a day in the sun and an hour or two of slow strolling. I can feel a good sort of tiredness creeping up on me, the sort that will enable me to sleep well tonight.

The menu contains all sorts of things I would like to order—especially a steak dish that I saw someone eating as we made our way through the restaurant to our table. But I settle, instead, for tuna shish kebab. I try to order fish, now, as much as possible when we are eating out. I stopped taking the fish oil tablets that I had read about as helpful for Crohn's disease in a nutrition book earlier that year; I couldn't stand the fact that I would be belching salmon for an hour or two after I took them, twice a day. So I try to get as much fish oil as I can naturally.

Unfortunately, I have never really liked fish all that much. But this dinner is delicious, and afterward I am happy and proud that I ordered and ate what I should have.

We make the long ride back to Worcester mostly in silence; Mike sits beside me in the passenger seat, and we talk occasionally about the upcoming school year. The girls are sleeping in their car seats, and Anne, in the far backseat, is nodding in and out herself.

Back at home, after depositing Mike at his apartment, we carry the girls into their beds and change them into their pajamas, still asleep. We pick up the toys they had managed to pull out before we had left that morning, and I am preparing to sit down and read or watch television in the family room.

"You know," Anne says to me, "I'm two weeks late at this point."

It has been nearly five months since my release from the hospital, and we have not yet had any luck with conceiving another child. Anne was initially concerned that perhaps my illness, or the medications I have been taking, have hurt her chances of becoming pregnant.

In the late spring we saw a fertility specialist. We explained my situation, and they arranged the appropriate tests for me. When we returned a week or two later, the news was good; I was still perfectly capable of impregnating Anne. That left more tests for Anne, which showed an irregularity in the way in which eggs were being produced in her cycle. She began taking a drug to regulate her cycle, and to receive an injection just prior to her most fertile time of the month.

But our summer travels had twice, in the last two months, prevented her from getting this injection at the right time, so we had essentially put our hopes on hold until we could return from our vacations and settle into our normal life in September.

"Do you have any pregnancy tests?"

"I have a two-pack upstairs."

I can tell she is reluctant, she doesn't want to find out for certain; if she doesn't take the test, she might still be pregnant. If she takes it and finds it negative, it means another month of waiting.

"Well, go take one!"

She heads upstairs, and I hear her opening the package and then closing the bathroom door. A minute later she comes downstairs.

"Well?"

"You have to wait two minutes," she says. "I can't look. You go up there and check."

So I go up there, but I don't go into the bathroom yet. She said it would take two minutes, so I go and sit on the bed. I have a momentary flash of all that a third child will bring us, good and bad: more diapers, more crying, more late nights and too-early mornings; more little tiny fingers, more smiles and laughter, more opportunities to help a small human being learn to negotiate the world.

Anne calls up the stairs to me.

"Well?"

"It hasn't been two minutes yet."

She walks back into the kitchen. I sit for another minute, and then take the short walk to the bathroom. Sitting on the counter I see the white plastic wand that holds the answer to our question.

I pick it up, and see it there, in the small window: a pink cross. Pregnant.

"Anne," I call downstairs. "Come up here and see for yourself."

She hurries upstairs and I show her. We embrace, and we both shed a few tears.

"Finally," I say.

"Finally," she says.

AND THAT MUST BE HOW this story ends, with the two of us once again locked in an embrace, once again with tears in our eyes, but in a situation so far removed from that embrace in the basement twelve months ago that I can hardly believe such extremes are possible in a single human life.

Those two embraces mark the opening and closing of a year that I have vowed never to forget. That first embrace carried me through seven months of illness and depression; this second embrace represents everything that has become valuable to me in the five months that have brought me to this instant in Anne's arms.

No one on this earth can predict how long this period of health and happiness will last for me, for us, but I banish that thought from my mind as soon as it enters.

Suspend this moment, frozen in time, and leave us here, now, locked in one another's arms, new life forming between us, and this moment—for just a moment—the only thing that matters in the world.

EPILOGUE

The Cardinal Rules for Crohn's Disease

The wisdom that I have gained over the course of the year described in this book is impossible for me to distill into an easy list of lessons. The lessons—about life with illness, about life with a physical self, about life as a human being—are too firmly embedded in the experiences I describe to rip them out of context and place them in a form that I think will reduce them to platitudes.

I do believe, though, that I learned some very specific lessons about the handling of chronic illness in general, and Crohn's disease in particular, and that those lessons might be usefully reiterated here for those readers who are struggling with their own chronic illnesses, or those of loved ones. I offer these especially to the newly diagnosed, with the hope that they can learn these lessons more easily and quickly than the year of illness it took me to absorb them into my body and my brain.

1. **Accept the presence of the disease in your life.** Do not expect God, or modern medicine, or any other substitute

for these entities, to cure you. Hope for a cure, pray for a cure, raise funds for a cure, campaign for public awareness for a cure—*but do not put your life on hold waiting for one.* Learn what accommodations you must make to the disease in your life; make those accommodations; get on with your life.

2. **Act quickly.** Learn the patterns of your disease, and become familiar with the signals that indicate an oncoming flare. When you can see those symptoms appearing, call your doctor *immediately*. Intervene as early as possible in order to prevent flares from spiraling out of control.

3. **Become your own patient advocate.** Not only can doctors not cure this disease; they cannot even yet tell us what causes it. Visit three different doctors and you might receive three different recommendations for treatment. Do not let doctors—or nurses, or dietitians, or your friends or families—make your medical decisions for you. Buy every book you can find, read as much scientific jargon as you can tolerate, and educate yourself about the human body and this disease. *Play an active role in your medical care.*

4. **Learn when to live and when to rest.** We are both blessed and cursed with a disease that comes and goes, and can spin us from remission into the hospital and back into remission again in the course of a few months. When you are healthy, *live intensely.* Gather sensation and experience as greedily as you can. When you are sick, rest and attend to your body. Although you may lose sight of this when you are in the middle of months or years of illness, a time will come again when you are healthy.

5. **Tell your story.** Tell it for yourself, first. We make sense of our lives through the stories we tell about ourselves. Telling

the story of your disease—on paper, on the Internet, to a friend—will help you see the disease's place in your life more clearly. Tell it for your friends and families, second. They need—and in most cases want—to understand what you are going through. Tell it, finally, for the rest of the world. Research funding and increasingly sophisticated medical care will increase with greater public awareness of the disease.

Resources for Crohn's Disease and Ulcerative Colitis

❧

My methods of gathering information about Crohn's disease were quirky and unsystematic, as most people's probably will be; usually when you most need information about the disease, you are in no position to expend the energy you would need to find it. Over the years, as I had more time to devote to my research, I have managed to put together a more comprehensive picture of those resources.

Most people, in their initial search for information, will turn to the World Wide Web. Be careful. The first sites that come up in some search engines are from people or organizations that are trying to sell you something. Do not be misled by any site promising you a complete cure for your disease—there is *no* cure for this disease at this time. Do not change your medications, or your diet, or your life without at least hearing the opinion of your doctor about whatever you have read.

But the Web can obviously be an excellent source of information as well, and it can provide you with opportunities to connect with others with Crohn's disease. Those connections—whether they take place in person, through support groups; online, through chat rooms; or on paper, through books—can be essential to maintaining your sanity and helping you see your own disease in perspective.

I offer here the top three resources that I turn to most frequently when I am in need of information, of help, and of support. These three resources are well-established and available at the time of publication, and I expect they will remain so as you are reading this book.

Crohn's and Colitis Foundation of America (CCFA)
386 Park Avenue South, 17th Floor
New York, NY 10016
Phone: 800-932-2423
Fax: 212-779-4098
Web: www.ccfa.org

Stop here first. The CCFA can provide you with information, with links to a plenitude of other resources, and with assistance in finding a physician. The CCFA funds research on the disease, and advocates for legislation to improve the quality of life for IBD sufferers. Join the organization to receive their newsletters, and donate to it at least a part of your budget earmarked for charity. Ten percent of the author's proceeds from this book will be donated to the CCFA.

Sklar, Jill, *The First Year: Crohn's Disease and Ulcerative Colitis*. New York: Marlowe, 2002.

I have found this book, authored by a medical writer with Crohn's disease, to provide the most easily accessible and up-to-date information on everything related to the disease: causes, symptoms, medications and their side effects, diet and nutrition, psychological issues. The book's table of contents provides easy access to information in just about every imaginable area related to inflammatory bowel disease.

IBD Sucks Web Forum: www.ibdsucks.org

Operated by Bill Robertson, this irreverently named Web site offers the opportunity to chat with fellow sufferers around the globe about everything related to IBD—and I mean everything. Those terrible and embarrassing parts of the disease that you thought you could never share with anyone? You'll find folks talking about that stuff here. In general, I have found the users of this site friendly and knowledgeable about the disease; posters often provide abstracts of current articles in medical journals, or links to other useful resources. A massive index directs you to conversation threads in literally dozens of specific areas. I go to this site when I'm feeling sorry for myself, and I usually feel better afterward.

. . . and, of course . . .

Learning Sickness Web Site: www.learningsickness.com

Read the comments of others who have read this book, post your own comments, and keep up with the author's progress and thoughts on chronic illness. The Web site also offers a fuller—and continually updated—list of links and Web resources.